D1385518

Colour of the powder

31	White or pale
32	Grey
33	Black
34	Yellow
35	Green
36	Rusty brown
37	Blue
38	Red
39	Orange, pink
40	Violet, purple

Fusi...

51	
52	Resistant to fusion
53	Non-fusible

Colour in a flame

54	Yellow
55	Green
56	Red
57	Orange
58	Violet
59	Yellow-orange
60	Yellow-green

e

69	Rare
	Common
70	Very common

Origin

71	In igneous rocks
72	In metamorphic rocks
73	In sedimentary rocks and hydrothermal deposits

Use

74	Metallurgy, building industry, etc
75	Jewellery
76	Chemical industries
77	Private collectors, scientific research

LUSTRE

41	Adamantine
42	Subadamantine
43	Vitreous
44	Metallic
45	Submetallic
46	Pearly
47	Silky
48	Resinous
49	Greasy, waxy, oily
50	Dull

Solubility

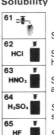

61	Soluble in water
62 HCl	Soluble in hydrochloric acid
63 HNO₃	Soluble in nitric acid
64 H₂SO₄	Soluble in sulphuric acid
65 HF	Soluble in hydrofluoric acid
66	Insoluble in acids

Polarised light with converging nicols

| 78 | Interference figure of dimetric crystals |
| 79 | Interference figure of trimetric crystals |

Preservation

| 80 ! | Perishable |

Minerals & Gemstones
an identification guide

G. Brocardo

English translation by Lucia Wildt

DAVID & CHARLES
Newton Abbot   London

CONTENTS

Cover illustration: *Wulfenite* Pb(MoO$_4$)
villa Ahumada, Sierra de Los Lamentos,
Chihuahua (Mexico)

Colour photographs: G. Brocardo
Pictograms: S. Pirlone

British Library Cataloguing in Publication Data

Brocardo, G.
  Minerals & gemstones.
  1. Precious stones—Identification
  I. Title   II. Minerali a colpo d'occhio. *English.*
  553.8   QE392

ISBN 0–7153–8283–7

© Priuli et Verlucca editori 1981
  David & Charles (Publishers) Limited 1982

Filmset by Keyspools Ltd, Golborne, Lancs
Printed in Italy

MINERALS AND THEIR IMPORTANCE

A mineral is any body which is natural, homogeneous, usually solid and inorganic.

'Natural' because it was formed, within a particular environment, through a completely natural process. Crystals obtained by man through water evaporation or gas sublimation, or through the solidification of molten matter, are not minerals but artifacts.

'Homogeneous' because its constituent particles (ions, atoms, molecules) periodically follow and repeat one another within the crystal structure.

'Solid' in so far as it has its own shape and volume, and its component particles are held together by great cohesive force. These particles can be arranged either in perfect order, thus producing crystals (*Fig* 1), or without any order, causing the amorphous state which we find, for instance, in the opal. Mercury and water, at normal temperature, are non-solid minerals.

Fig 1 Crystal structure of
sodium chloride

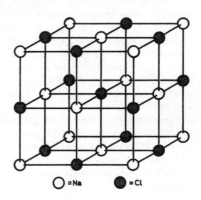

○ = Na ● = Cl

'Inorganic': overall minerals are not the product of organic processes or of living creatures, but are almost always derived from inorganic matter.

The fossils (eg coal) with a vegetal origin are not real minerals and are studied only as a marginal branch of mineralogy. Sometimes, a vegetal matter can undergo such a transformation as to crystallize, in which case the end product can be considered a mineral. This is the case, for instance, of limestone, which originates in the shells of foraminifers, and of jasper, originally formed by the shells of diatoms.

Rocks, on the other hand, are conglomerates of various minerals, all forming an integral part of the lithosphere. Rocks are therefore heterogeneous and widespread. In some cases one single mineral can be so widespread as to be considered a rock: its spread and relative heterogeneity cause it to be called 'simple rock'. The limestones of the Alpi Apuane (Tuscany, Italy) or the rock-salt of Galicia (Poland) are such simple rocks. Well-crystallized and developed minerals are rare.

Minerals which are found in large deposits can be exploited and assume a particular interest at various levels:

1. Economical: they provide the raw materials for industry. Rich mineral deposits are an important factor in a nation's economical power. The fabulous jewels which have been found throughout the world are also minerals: rubies and diamonds, emeralds and topazes, fiery opals and limpid rubellites. Native gold is the basis of international currencies. Minerals are also used in optics, in the construction of sophisticated scientific instruments, and in the building and ceramics industries.

2. As a science, mineralogy is a fascinating world, with many avenues of research, which have all produced, and still lead to, satisfying results. Indeed, minerals provide us with ample scope for research: their chemical composition, their lattice structure, the crystallographic laws which determine their development, their genesis, their physical characteristics.

3. With the spreading of knowledge at all levels, many are the amateurs who get passionately involved in the search for, and the study of, minerals. The crystals' beauty, shape, colour, light, are all attractive. Rarity often confers a real value on the mineral and causes a so-called 'fever'. Popular mineralogy has even provided a stimulus for university institutes, who are increasingly involved in the definition of mineral species and often find that specimens discovered by the amateur are indeed a new form. By the same token, private collections are growing alongside those of universities and museums, and, although smaller, they often prove extremely interesting for their topographical value and the presence, among the specimens they contain, of rare and entirely new species. These new species are caused by the association between the atoms of the chemical elements, present in the lithosphere, which produce new molecules within the crystal structure.

Let us have a look at the percentage, by weight, of the chemical elements present in the lithosphere:

Element	Symbol	Percentage	Element	Symbol	Percentage
Oxygen	O	46.6%	Beryllium	Be	0.0007%
Silicon	Si	27.7%	Samarium	Sm	0.0006%
Aluminium	Al	8.1%	Gadolinium	Gd	0.0006%
Iron	Fe	5.0%	Praseodymium	Pr	0.0005%
Calcium	Ca	3.6%	Scandium	Sc	0.0005%
Sodium	Na	2.8%	Arsenic	As	0.0005%
Potassium	K	2.6%	Hafnium	Hf	0.0004%
Magnesium	Mg	2.1%	Dysprosium	Dy	0.0004%
Titanium	Ti	0.4%	Uranium	U	0.0004%
Hydrogen	H	0.1%	Boron	B	0.0003%
Phosphorus	P	0.1%	Ytterbium	Yb	0.0003%
Manganese	Mn	0.1%	Erbium	Er	0.0002%
Sulphur	S	0.05%	Tantalum	Ta	0.0002%
Carbon	C	0.03%	Bromine	Br	0.0002%
Chlorine	Cl	0.03%	Holmium	Ho	0.0001%
Rubidium	Rb	0.03%	Europium	Eu	0.0001%
Fluorine	F	0.03%	Antimony	Sb	0.0001%
Strontium	Sr	0.03%	Terbium	Tb	0.00009%
Barium	Ba	0.02%	Lutetium	Lu	0.00008%
Zirconium	Zr	0.02%	Thalium	Tl	0.00006%
Chromium	Cr	0.02%	Mercury	Hs	0.00005%
Vanadium	V	0.01%	Iodine	I	0.00003%
Zinc	Zn	0.01%	Bismuth	Bi	0.00002%
Nickel	Ni	0.008%	Thulium	Tm	0.00002%
Copper	Cu	0.007%	Cadmium	Cd	0.00001%
Tungsten	W	0.007%	Silver	Ag	0.00001%
Lithium	Li	0.006%	Indium	In	0.00001%
Nitrogen	N	0.005%	Selenium	Se	0.00001%
Cerium	Ce	0.005%	Argon	Ar	0.000004%
Tin	Sn	0.004%	Palladium	Pd	0.000001%
Yttrium	Y	0.003%	Platinum	Pt	0.0000005%
Neodymium	Nd	0.002%	Gold	Au	0.0000005%
Niobium	Nb	0.002%	Helium	He	0.0000003%
Cobalt	Co	0.002%	Tellurium	Te	0.0000002%
Lanthanum	La	0.002%	Rhodium	Rh	0.0000001%
Lead	Pb	0.002%	Rhenium	Re	0.0000001%
Gallium	Ga	0.001%	Iridium	Ir	0.0000001%
Molybdenum	Mo	0.001%	Osmium	Os	0.0000001%
Thorium	Th	0.001%	Ruthenium	Ru	0.0000001%
Caesium	Cs	0.0007%	Radium	Ra	0.0000000001%
Germanium	Ge	0.0007%			

IN SEARCH OF MINERALS

Beginners could do worse than explore the dumps of both active and disused mines, particularly since the latter are usually marked on maps and therefore easily traceable. It is a help to know which mineral is, or was, mined in any given place: any mineralogy text will then show which other minerals usually accompany that particular one and the collector will thus be able to derive greater profit from a visit to the dump. Minerals are often accompanied by the gangue, ie the rock which contains them; when exploring a dump, the collector should therefore pay particular attention to the varieties of gangue in order to find the sought-after mineral. Any piece which appears to be of interest should be carefully wrapped up and examined later at ease.

Other useful places are caves, where rocks can easily be found. Here again, it is helpful to know which mineral usually accompanies which rock, as a visit to a cave will then be that much more profitable.

The more experienced collector usually conducts his research in high or mountainous areas. Geological maps tell him which rocks appear on the surface of the soil and which minerals they contain. Many rocks have no minerals of any interest, others can hide real jewels in their fissures and cavities. In order to be fairly successful it is essential to have prepared oneself properly by studying maps and books. The search can then be carried out on clear rocks, like granite, granodiorite, gneiss; it is here that one can find quartz, albite, adularia and often even some marvellous calcite crystals. Less frequent, but still presenting good crystal forms, are the fluorites, apatites and the glittering haematites.

Syenites could contain epidote, titanite, apatite, prehnite, axinite and tremolite. Metamorphic rocks or crystal schists can disclose anatase, brookite and rutile. Amphibolites are often accompanied by splendid epidotes, adularia, albite, tremolite and titanite. In all these rocks one can also find zeolites as well as rarer specimens, such as bazites and monazites.

Much can be achieved by exploring crystal limestones and dolomitic rocks; it is here that one can often find beautiful calcites, quartz dolomite, rutile, tourmaline, and metallic sulphides like blende, galena, tennantite and pyrite. One can also come across, although more rarely, anidride, tungstenite, corundum, scapolite, flogopite, taramelite, wenkite and celsian. Another area extremely rich in mineral species is the border area between limestones, dolomites and igneous rocks; here one can find extremely beautiful vesuvianites, grossulars, pyroxenes, spinels, zoisite and ghelenite. Even serpentine rocks are extremely interesting, particularly where they come into contact with rodingites and grenatites; it is here that one can find splendid grossulars, vesuvianites, multicoloured andradites, diopside, epidotes, clinochlores, magnetites, and, less frequently, zoisite, apatite, titanite, clinozoisite. Rarer, however, are the perovskites, uvarovites, corundums and zircons. Sometimes these serpentine rocks contain veins rich in garnets and magnetite; and here the action of water can produce very interesting minerals, such as artinite, brugantellite, hydromagnesite, nesquehonite, auropyrite and brucite. Talc schists and chlorine schists often contain magnesite, apatite and actinolite. Finally, such minerals as andalusite, kyanite, sillimanite, staurolite, almandine and hornblende are often found in metamorphic rocks.

Igneous rocks are an ideal source of minerals. Melaphyre, porphyry and basalt all have cavities covered in splendid crystals, usually of analcyme, apophyllite, strontianite, augite, spinels, quartz, celestite and above all zeolites (heulandite, mordenite, cabasite, stilbite, harmotome, natrolite).

Obviously, if one is to be successful in the search for minerals, a good knowledge of rocks is indispensable.

HOW MINERALS ARE FORMED

The origin of minerals is closely connected to the formation and evolution of rocks. The latter are formed either by the solidification of the magma, by sedimentation of organic deposits, or by the metamorphosis of existing rocks.

a) MINERALS WITH MAGMATIC ORIGIN

Magma is a molten mass with very high temperatures and comprising various elements, many of them volatile. Usually the prevailing constituents are silicon, aluminium, calcium, magnesium, sodium, potassium and iron. A large quantity of water, in the form of steam, is also present. The loss of heat produces the crystallization of the magma, which does not occur homogeneously but according to a selective and gradual procedure. The various stages of this crystallization are deeply affected by pressure and temperature and are called orthomagmatic, pegmatitic, pneumatolytic and hydro-thermal (*Fig* 2).

Fig 2   Magmatic basin

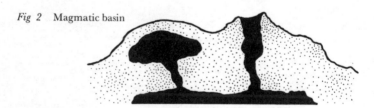

ORTHOMAGMATIC STAGE

The first to crystallize are the accessory minerals: zircon, titanite and apatite. These are followed by those minerals containing iron and magnesium, such as pyroxene, amphibole, biotite. Later on come albite, anortite, orthoclase and quartz. Also appearing at this stage are some metalliferous minerals such as chromite, pyrrhotine, magnetite and titanite.

The nearer the magma is to the surface the quicker the cooling-down process: this considerably affects the order of crystallization. During this stage, the volatile elements such as chlorine, flourine, boron, steam, have little part in the crystallization process and concentrate within the residual magma, thus increasing its internal pressure.

PEGMATITIC STAGE

This takes place at a lower temperature (600–700°C) but at higher pressure than the preceding stage; the volatile elements are now taking full part. It is during this stage that large crystals are formed. Collectors are particularly interested in these pegmatites, both for their beautiful shapes and the rarity of the mineral species themselves. They include precious berils (emeralds, aquamarines), topaz, many varieties of tourmaline, apatites, spodumenes, uraninites and other rare species.

During this stage, all residual magma crystallizes. Slowly the temperature drops to 500–372°C. Not all volatile elements have taken part in this stage: the remaining ones now concentrate in the fissures left by the solidification of the magma or in those of the surrounding rocks.

PNEUMATOLYTIC STAGE

The leftover gases and the steam react with the nearby rocks and produce further mineral species. This stage is characterized by cassiterite, molybdenite, scheelite, vesuvianite, ilvaite, grossular, spessartite and other garnets. The mineral deposits thus formed are called 'pneumatolytic contact deposits' since they formed in the contact area between the magma and the surrounding rocks.

HYDROTHERMAL STAGE

When the temperature of the crystallized magma drops below 372°C, steam liquefies and deposits its mineral constituents in fissures and fractures, forming veins of precious minerals which partly replace limestones. Many sulphides crystallize in this way: blende ZnS, pyrite FeS_2, galena PbS, antimonite Sb_2S_3, cinnabar HgS, fluorite CaF_2, siderite $FeCo_3$, baritine $BaSO_4$, quartz SiO_2. Some species are typical of high, medium or low temperatures which form according to the distance from the magma.

If the magma, rather than solidifying at a certain depth, rises towards the surface, many gases and vapours sublimate along the rocky surfaces which line the fissures. Thus new minerals are born, among which are sulphur, realgar AsS, orpiment As_2S_3, ematite Fe_2O_3, tenorite CuO, atacamite $CuCl_2.3Cu(OH)_2$ and others.

b) MINERALS OF SEDIMENTARY ORIGIN

Rocks are eroded and transformed by the physical and chemical action of the atmosphere. The minerals they contain are dissolved in water and carried away to other sites. As they settle again, or re-crystallize following the evaporation of water, they form the secondary deposits, of which those due to chemical or biological processes are particularly important.

Calcium carbonate, $CaCO_3$, which is insoluble, is turned to calcium bicarbonate $Ca(HCO_3)_2$, which is soluble, by carbon dioxide

CO_2 and water H_2O. Limestones thus slowly dissolve and water carries them along soil fissures and water courses as far as the sea. During the journey, as water evaporates, calcium carbonate forms again in the shape of stalactites and stalagmites, or of travertine or fine calcite crystals. The same happens to dolomites, although they are less soluble, and to chalk and anhydrite, relatively soluble, which re-crystallize as water evaporates. Deposits of this kind are often huge.

Less frequent are deposits of potassium salts caused by silvite, carnallite, kainite, bischophyte, leonite, etc. They all owe their origin to chemical processes. It seems that sulphur may have biological origins due to the action of bacteria on chalk. And limonite, or 'swamp iron' is also derived from bacteria acting on iron and settling on the bottom of swamps and lakes.

Many minerals formed by the alteration of sulphurous deposits are also considered to have a sedimentary origin. This alteration produces sulphuric acid, H_2SO_4, which reacts with other minerals. If the deposits are in contact with air (when they lay on the surface), oxidation and carbonization also take place, This is the so-called 'oxidation area' of mineral deposits, within which many species are found, such as siderite, limonite, cerussite, anglesite, smithsonite, hemimorphite, azurite, cuprite and chrysocolla. Part of the solutions created by the alteration of superficial deposits can descend to some depth, far from the atmosphere, and leave the salts in an environment which is no longer oxidizing but 'reducing'. Here can be found native elements such as silver and copper, and other species such as covelline, calcosine, etc.

c) MINERALS OF METAMORPHIC ORIGIN

Rocks are not static objects: they go through various alterations due to dislocations, sedimentation and changes of depth. This causes tensions, high pressures and high temperatures, all of which determine physical and chemical alterations of the mineral components.

As far as depth is concerned, there are three characteristic environments:

1. The area between 5000 and 7000m
2. The middle area between 7000 and 12000m
3. The deepest area between 12000 and 20000m

In the top area temperature is low (200°C) and high pressures are uneven; here find their origin, as derivatives of other minerals, many of the silicates, such as chlorite, sericite, tremolite, talc, actinote, albite, epidote and titanite. In the middle area temperatures reach 400°C, pressure is still uneven, and minerals such as staurolite, cyanite, garnets, hornblende, muscovite and biotite are formed. In the lowest area, at much deeper levels, temperatures reach 600°C, pressure is high and more or less even over the mass surface. The minerals which are formed in this environment are similar to those originated by the solidification of the magma: sillimanite, cordierite, graphite, almandite garnet, orthoclase, plagioclase, olivine and pyroxene.

There are three methods normally used to identify one mineral species:

1. Recording the geometrical appearance of the crystals and reconstructing the shape and system of crystallization
2. Analysing the physical characteristics of the sample
3. Carrying out certain chemical experiments, as long as they are not harmful to the mineral

One can also use extremely expensive instruments which exploit X-rays and microprobes, but this requires well-equipped and specialised laboratories.

1. GEOMETRICAL STRUCTURE: BASIC CRYSTALLOGRAPHY

'Crystals' are minerals with original, polyhedral surfaces which are the result of the regular array of the molecules forming the lattice; they are eminently symmetrical bodies whose elements are physically equivalent faces, edges and vertices (*Fig* 3). A crystal can be divided by a 'mirror plane' into two parts which are mirror images of each other; such a plane is also called a 'plane of symmetry' (*Fig* 4). Crystals can also be made to rotate around one or more axes to show the spatial

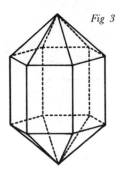

Fig 3   Quartz crystal: a symmetrical body

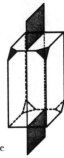

Fig 4   Symmetry plane

superimposition of their equivalent elements; the axes are called axes of symmetry. Some crystals possess an internal point from which physically equivalent axes depart: this is the centre of symmetry.

Axes are called binary, ternary, quaternary, senary, according to the degree of rotation (ie 180°, 120°, 90°, 60°) required to obtain the

11

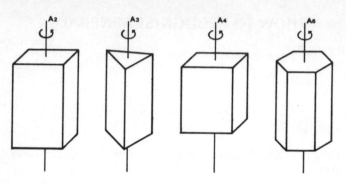

Fig 5 Symmetry axes: A_2 = binary; A_3 = ternary; A_4 = quaternary; A_6 = senary

superimposition of equivalent elements (*Fig 5*). The faces of a crystal are not important for their size but for their inclination: natural crystals are not perfect solids, as some faces are more or less developed than others. Nicola Stenone first formulated a law based on the dihedral angles determined by two faces: 'In all crystals belonging to the same chemical group and the same stage, the dihedral angles of two homologous faces have a constant value, as long as there are no temperature variations.' If crystals are large enough, the dihedral angles can be measured with an application goniometer: Abbot Hauy obtained thus some amazingly exact results (*Fig 6*).

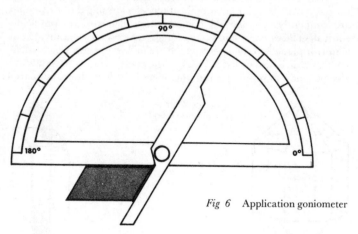

Fig 6 Application goniometer

Perfect measurements of dihedral angles can be obtained with reflection goniometers (*Fig 7*).

The crystal is placed on the graduated plate of the goniometer; a beam of light is then projected onto one face through a collimator; the plate is then rotated and the reflection of the light observed through a telescope and its value read on the fixed index expressed by the graduated plate. The plate is then rotated again: the light will

Fig 7   Reflection goniometer

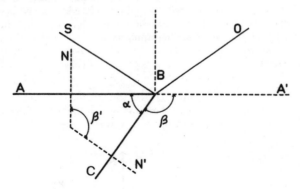

disappear and reappear when another face has taken the place of the previous one. The rotation value is then read again; it will not give the dihedral angle between the two faces but the supplementary angle, which is the same as the angle formed by the perpendiculars to the two faces in question (*Fig* 8).

Fig 8   AB and BC = crystal faces determining the dihedral angle: β is the angle measured by the goniometer and is equal to β', which is given by the normals to the two faces, N and N'

Useful data are acquired by comparing the values of the dihedral angle of a given crystal with that of other well-known specimens.

What is most interesting to a crystallographer is to identify the forms of crystals, to trace them back to their system and crystallization class and thereby be able to identify the mineral with optimum precision. 'Form' is the group of physically equivalent faces, mutually bound by symmetrical elements. Many forms can occur in a crystal (*Fig* 9).

A cube for instance is made of six faces which are symmetrically equivalent; it is a simple form, enclosing a certain amount of space and

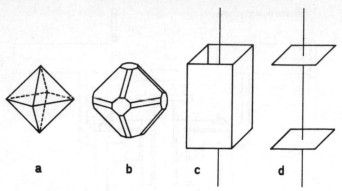

Fig 9   Crystal shapes: a = octahedron (simple shape, closed); b = a shape formed by a cube and octahedron, a rhombododecahedron; c = tetragonal prism (simple shape, open); d = pinacoid

therefore called 'closed'. Prisms, on the other hand, do not enclose space and are therefore 'open' forms (*Fig 9*); in order to exist as natural forms they need a second form to close them. This second form is the 'pinacoid', ie a pair of parallel and equivalent faces.

When studying crystal forms it is very important to define the position of the faces. This is done by choosing three reference axes in agreement with the crystal and three of its edges not lying on the same plane. The lines which, parallel to the edges, cross a point inside the crystal form the 'crystallographic axes' which, all together, give us the 'axial cross'. The axis facing the observer is axis x; the horizontal one, parallel to the observer, is axis y; the vertical one is axis z. The positive directions of each axis are:

for x from the centre of origin towards the observer
for y from the centre of origin towards the right of the observer
for z from the centre of origin upwards (*Fig 10*)

Fig 10

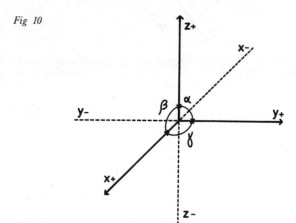

The positive directions of each axis form angles defined as follows:

$\alpha = yz$
$\beta = xz$
$\gamma = xy$

In general $\alpha = \beta = \gamma = 90°$. Alpha, beta and gamma (α, β, γ) are called the 'angular constants' of the crystal.

Once the three axes have been defined, a face is chosen which crosses the three axes on the positive side; this is called the fundamental face, as the position of the others is related to it. This face cuts the axes into three segments called 'parameters'; they are OA, OB, OC, respectively indicated as a, b, and c. It is possible to establish the relationship between the parameters as a:b:c, which is called the 'parametric ratio' of the fundamental face. The parametric ratio and the angular constants form the crystallographic constants of a crystal. The position of the other faces is established by working out the relationship between the parameters of the fundamental face and those of the others occurring on the same axes (*Fig* 11):

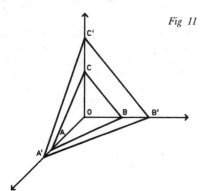

Fig 11

$$\frac{OA}{OA'} \quad \frac{OB}{OB'} \quad \frac{OC}{OC'}$$

These ratios are expressed by three numbers which are indicated by the letters h (for axis x), k (for axis y) and l (for axis z). These numbers are called 'indexes' of the face. Thus we have:

$$\frac{OA}{OA'} = h, \quad \frac{OB}{OB'} = k, \quad \frac{OC}{OC'} = l$$

Experiments have shown that the indexes of the faces of both natural and artificial crystals are usually small ($h + k + l < 10$).

Indexes are usually expressed by numbers of letters. If, for instance, the fundamental face, in cutting across the crystallographic axes x,y,z, produces parametric values which are respectively 0.384, 0.680 and 1.010, and if the other face produces the parameters 0.768, 1.360 and ∞ (infinity), we shall have the following indexes:

for x $\dfrac{0.384}{0.768}$; for y $\dfrac{0.680}{1.360}$; and for z $\dfrac{1.010}{\infty}$

that is $\frac{1}{2} \frac{1}{2} 0$ which can also be written as (1 1 0)

If, on the other hand, the other face produced, on the three axes, the parameters 0.384, 1.360 and 2.020, the ratios will be as follows:

for x $\dfrac{0.384}{0.384}$; for y $\dfrac{0.680}{1.360}$; and for z $\dfrac{1.010}{2.020}$

hence h:k:l = 1:$\frac{1}{2}$:$\frac{1}{2}$ or (2 1 1)

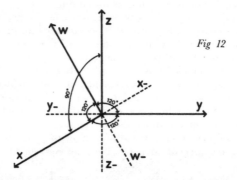

Fig 12

In the case of certain crystals it is easier to use four crystallographic axes: x,y,w on the same plane at an angle of 120° between each, and z, orthogonal to the former (*Fig* 12). x is positive towards the observer, y at 120° to the right; and at 120° from y, the positive w. The parameters of the fundamental face will therefore be four: a:a:a:c (ie the unit of measurement on the three complanate axes is the same, and it is different on z), and four will be the indexes of each face: h k f l. The result, known as Hauy's law, is that the indexes of crystal faces are represented by rational and unusually low numbers. Hauy reached this formulation basing his ideas on his own conception of crystal structure, ie the periodical and three-dimensional repetition of the smallest solid. All experimental observations confirm his theory. Nowadays one no longer refers to these solids, but to atoms, ions and molecules as constituents of a crystal; but their spatial distribution is just as Hauy has seen it (*Fig 13*).

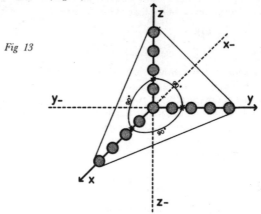

Fig 13

Crystals possess varying symmetries; for instance, chalk has one reflection plane, one binary axis and the centre of symmetry; rock-salt has nine reflection planes, four ternary axes, three quaternary axes, six binary axes and the symmetry centre; albeit only the symmetry centre. It has been demonstrated that there are 32 possible groupings of the symmetry elements and these determine the various crystal classes: they express the possible symmetry of any crystal, whether natural or artificial.

The symmetry degree indicates the sum of the various elements (planes, axes and centre) which are typical of each class. The crystallographic constants of each class provide the basis for their grouping into systems, each of which includes the classes which have the following elements in common: the same type of axial cross and the same parametric ratio. Crystal forms, in each class, are indicated by placing the indexes within two double brackets; the faces are shown by their indexes within round brackets. The indexes of the fundamental face are the ratio between its parameters, therefore its indexes will be (1 1 1). It follows that the form is indicated by the formula (111) and the fundamental face by (111).

Seventeen systems are in turn subdivided into three groups according to the parametric ratio of their fundamental face. These groups are:

1. *Monometric*—composed exclusively by the homonimous system, also called cubic, with a fundamental face which determines three equal parameters on the crystallographic axes

2. *Dimetric*—formed by hexagonal, trigonal and tetragonal systems where the fundamental face produces two different values on the crystallographic axes: equal values on complanate axes, different on z

3. *Trimetric*—formed by the rhombic, monoclinic, triclinic systems, where the fundamental face produces three different values on the crystallographic axes

Specialised texts explain crystallography in great detail as a good knowledge of it is essential to recognise minerals; for instance, an epidote can easily be differentiated from a vesuvianite because the latter has a tetragonal habit and the former a monoclinic one.

NB The two pages which follow give a detailed table of crystal systems

Table of crystal systems

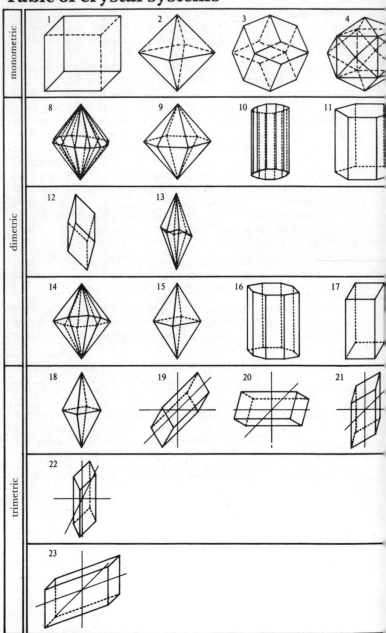

monometric

dimetric

trimetric

18

Forms of crystal systems

1	cube or hexahedron	(100)
2	octahedron	(111)
3	rhombic dodecahedron	(110)
4	tetracisexhahedron	(hkO)
5	icositetrahedron	(hkk)
6	triacisoctahedron	(hhl)
7	hexacisoctahedron	(hkl)
8	dihexagonal bipyramid	
9	hexagonal bipyramid	
10	dihexagonal prism	
11	hexagonal prism	
12	rhombohedron	
13	scalenohedron	
14	ditetragonal bipyramid	(hkl)
15	tetragonal bipyramid	(hhl)
16	ditetragonal prism	(hkO)
17	tetragonal prism	
18	rhombic bipyramid	(hkl)
19		(Okl)
20	rhombic prisms	(hOl)
21		(hkO)
22	monoclinic prism	(hkO)
23	pinacoid	

	triclinic	monoclinic	rhombic	tetragonal	trigonal	hexagonal	cubic
systems	triclinic	monoclinic	rhombic	tetragonal	trigonal	hexagonal	cubic
crystallographic constants	$\alpha \neq \beta \neq \gamma \neq 90°$ a:b:c	$\alpha = \gamma = 90°$ $\beta \neq 90°$ a:b:c	$\alpha = \beta = \gamma = 90°$ a:b:c	$\alpha = \beta = \gamma = 90°$ a:a:c	x,y,w complanate to 120° z orthogonal a:a:a:c	α,y,w complanate to 120° z orthogonal a:a:a:c	$\alpha = \beta = \gamma = 90°$ a:a:a
number of classes	2	3	3	7	7	5	5
most symmetrical class	pinacoidal C	prismatic P, A_2	bipyramidical rhombic 3P, $3A_2$, C	bipyramidical ditetragonal 5P, A_4, $4A_2$, C	scalenohedric ditrigonal 3P, A_3, $3A_2$, C	bipyramidical dihexagonal 7P, A_6, $6A_2$, C	hexacisoctahedric 9P, $3A_4$, $4A_3$, $6A_2$, C
type			A_2	A_4	A_3	A_6	$4A_3$

19

LATTICE STRUCTURE OF CRYSTALS

A crystal in the solid state is a three-dimensional arrangement of particles in a definite repeating pattern. These particles (atoms, molecules or ions) occupy the vertices of minute parallelepipeda which, bordering one another top and bottom or side by side, form what is known as the *space lattice* (*Fig* 14). The particles at the vertices of the parallelepipeda, or 'networks', are the 'nodes', or lattice-points of the structure. A series of such points lying in the same direction makes a 'row'; several rows in the same direction define a 'lattice plane'.

Fig 14 Lattice structure

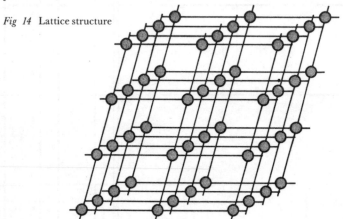

There exist in all 14 elementary cell-types which, superimposed or otherwise inter-related, cover every possible combination of points in space (*Fig* 15).

To describe their patterning we may introduce a new symmetrical quantity, absent from crystalline solids considered as finite bodies: the *translation vector*. From a given point on a specimen this will extend it to infinity in a given direction and at constant intervals.

$\vec{\tau}$ is the new symmetrical quantity ie the translation vector.

The crystal, seen as a vast collection of atoms (or ions, or molecules) will thus disclose new elements of symmetry, as mirror planes and rotation axes can combine with the translation vectors to exhibit, respectively, gliding planes and spirals.

There are 230 possible combinations, known as *space groups*, of the internal symmetrical elements of a crystal.

The lattice structure of crystals was demonstrated in a brilliant experiment by the Alsatian physicist Max von Laue, together with the X-ray scientists Friedrich and Knipping, in 1912. At his suggestion, the two specialists passed X-rays through a crystal of blende. The diffraction observed as the rays passed through the lattice-planes not only proved the wave-like nature of the rays, it also revealed the discontinuous and lattice-like nature of crystals themselves. The Laue codification of crystal symmetry is of the greatest use in the study of crystal structure.

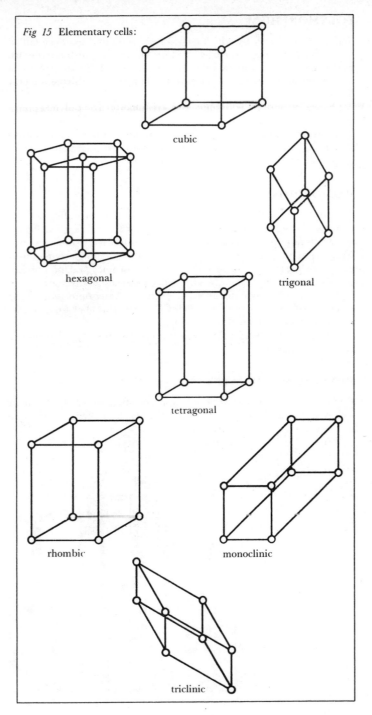

Fig 15 Elementary cells:

cubic

hexagonal

trigonal

tetragonal

rhombic

monoclinic

triclinic

POLYMORPHISM

This is the property by which a chemical compound may crystallize into differing lattice structures. The system or class of the structure changes, while the mineral is chemically unaltered. Polymorphism is due to certain conditions of pressure and temperature during mineral formation.

Quartz α and β, tridymite and cristobalite are polymorphous derivatives of silicon oxide.

Typical examples of dimorphism are pyrites (cubic) and marcasite (rhombic); diamond (cubic) and graphite (hexagonal); calcite (trigonal) and aragonite (rhombic).

ISOMORPHISM

This is the property by which minerals with some differences of composition will crystallize into identical forms; that is, their lattice vertices retain the same pattern but are tenanted by dissimilar atoms. Many mineral species of analogous composition, having atoms or ions of approximately the same radius, are isomorphic, and isomorphic substitution, though sometimes limited, can occur on any scale.

The trigonal carbonates afford an example of an isomorphic series: calcite ($CaCO_3$), rhodocrosite ($MnCO_3$), siderite ($FeCO_3$), smithsonite ($ZnCO_3$), magnesite ($MgCO_3$).

Isomorphism is very common in nature, unique crystal structure being exceptional.

CRYSTAL AGGREGATES

Crystals in the natural state occur more often in groups than in isolation. Clusters growing from a flat base are called *druses*; those growing from a concave surface are *geodes*.

Where there is iso-orientation, the crystals will form 'parallel groups' (*Fig* 16).

Fig 16 Parallel groupings

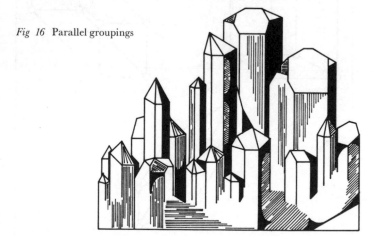

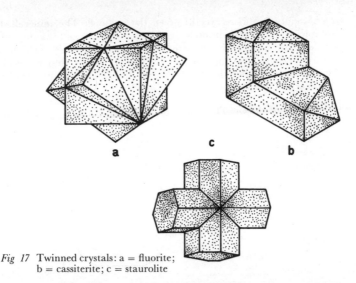

Fig 17 Twinned crystals: a = fluorite;
   b = cassiterite; c = staurolite

Another important phenomenon is that of *twinning* (*Fig* 17) which occurs when two or more crystals develop in obedience to definite twinning laws. Crystals will also assume typical shapes, classified as radiated, fibrous, laminated, and so on. There are also multiple forms such as stalactitic, stalagmatic, mamillated, coralloid, etc.

2. PHYSICAL PROPERTIES OF MINERALS

Suitable specimens from which to observe crystal structure and identify a mineral are not always available; often we have only fragments, or powder, to examine and must rely on other data. A study of the physical properties of minerals will tell us a great deal and can be undertaken with a few inexpensive pieces of equipment, though there are of course some experiments which demand costly apparatus and specialized research training.

The physical properties of crystals are of two kinds—scalar and vector. The former are quantities independent of direction; the latter vary according to direction.

Specific gravity and fusibility are scalar properties.

SPECIFIC GRAVITY

This is one of the most important properties of a mineral and is defined as the ratio of its weight to that of an equal volume of distilled water at 4°C. The denser the atoms in a structure, the greater its specific gravity, which can be measured with a pyknometer, a Westphal balance, or with a torsion balance and heavy liquids.

The pyknometer, or density bottle (*Fig* 18), is a small, wide-necked flask, having a ground-glass stopper down the centre of which runs a capillary hole. The mineral is accurately weighed, and this weight is given the value M. The density bottle is then filled with the distilled water, weighed, and its weight given the value P. The mineral, if insoluble in water, is then introduced into the bottle and the stopper replaced, care being taken to ensure that the water reaches no higher than its previous level in the capillary hole. Bottle and mineral, weighed together, give the weight N and the specific gravity may be expressed as:

$$\frac{M}{(P+M)-N}$$

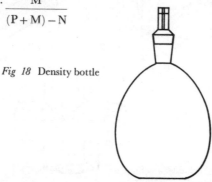

Fig 18 Density bottle

Specific gravity, being a useful diagnostic indication, is always included in a mineral description. To measure it with a Westphal balance (Fig 19) the mineral specimen is first weighed on the platform, out of water, and the balance trimmed with rider weights. It is then removed from the platform, known weights are substituted and the balance is re-adjusted. We thus have the weight of the specimen, which is then introduced, on the pan, into the water. The balance now loses its equilibrium for the specimen receives an upthrust equal to the weight of water displaced. We now find this weight by adjusting the

Fig 19 Westphal balance

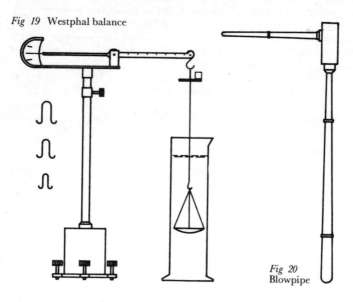

Fig 20
Blowpipe

balance, and the ratio between this weight and that of the specimen prior to immersion gives the specific gravity.

If a mineral is soluble in water we must employ non-solvent, or heavy, liquids and a torsion balance. In practice, observations of specific gravity can be made directly from nature. Denser materials, the residue of rock-debris, will usually be noticed upstream, while the lighter sorts have been brought downstream by the action of flowing water.

Minerals are graded as follows, according to their specific gravity:

very light, with a specific gravity of 2
light, with a specific gravity of from 2 to 3
dense, with a specific gravity of from 3 to 5
very dense, with a specific gravity of from 5 to 10
densest, with a specific gravity of 10 and upwards

FUSIBILITY

The melting-point of minerals is another scalar property very useful to know but difficult to calculate. We can, however, determine whether or not a mineral is fusible by experimenting with small chips in the flame of a Bunsen burner, which exceeds 1000°C, or with a blowpipe (*Fig* 20) and charcoal. We also have von Kobell's empirical scale of 7 standard minerals for purposes of comparison; it includes the following typical minerals:

1. Antimonite—melting-point 525°C in flame of match
2. Natrolite   —melting-point 800°C in gas flame
3. Almandine —melting-point 1050°C in gas flame
4. Actinolite —melting-point 1200°C in flame of blowpipe; when the fused mineral will curl at the edges
5. Orthoclase —melting-point 1300°C in flame of blowpipe; does not readily curl at the edges
6. Bronzite   —melting-point 1400°C in flame of blowpipe; curls if in thin splinters
7. Quartz   —melting-point above 1400°C; will not melt in flame of blowpipe

In practice many species can be recognised by means of fusibility tests, and most descriptions either give the melting-point or indicate whether or not a mineral is easily fusible, if at all.

Vector properties are those of hardness and cleavage, the electrical and magnetic qualities, and optical behaviour.

HARDNESS

This is defined as the degree of resistance of a crystal to scratching.

Tests with a steel needle will show the degree of hardness of a sample and tell us whether it is, for example, calcite, apatite or quartz. The hardness of a mineral depends on its cohesion, and therefore on its structure, and in all crystals, even those belonging to the monometric, or cubic, system, hardness varies with structural direction. Though difficult to assess exactly, hardness can be measured with various types of sclerometer.

On Seebeck's sclerometer (*Fig* 21) the specimen is held on a moving platform above which is mounted a needle with weights attached. The degree of hardness is gauged by the weight required to produce a minimal scratch. The Mohs empirical scale is used as a standard of hardness. Variations of hardness between the 10 graded minerals of this scale are irregular, but each mineral will scratch those above it.

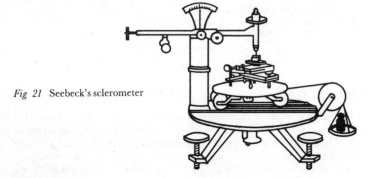

Fig 21 Seebeck's sclerometer

Mohs scale

1. Talc	very soft (1–2)	can be scratched with the fingernail
2. Gypsum		
3. Calcite	soft (2–3)	can be scratched with an iron wire
4. Fluorite	semi-hard ($3\frac{1}{2}$–$4\frac{1}{2}$)	can be scratched with a penknife
5. Apatite	hard (5–$6\frac{1}{2}$)	not easily scratched with a penknife
6. Orthoclase		
7. Quartz		
8. Topaz	very hard ($6\frac{1}{2}$–10)	cannot be scratched with a steel point
9. Corundum		
10. Diamond		

If a specimen will scratch one of these standard minerals but not the one below, its hardness is reckoned as being halfway between the two grades.

CLEAVAGE

Most, though not all, crystals have the property of shearing along certain planes, known as cleavage planes.

Calcite breaks easily; pyrites, magnetite, quartz and granite do not break. Cleavage is strictly related to crystalline structure. It occurs between planes whose bond-distances are long and therefore weak, and cleavage planes are always parallel to an actual or possible face of the crystal. Malformed crystals can be reduced in size, and regain their perfect forms, by cleavage.

Many crystals will split easily along one or more planes and the angle between planes of cleavage may help in identification. Thus pyroxenes, which split along almost perpendicular planes, are distinguished from amphiboles, splitting alone planes that form angles of about 120° and 60°. Metals are too compact to cleave at all.

ELECTRICITY

Native metals, sulphides and oxides, are conductors of electricity. Most minerals, on the other hand, are 'dielectric', ie non-conductors of electricity.

Some dielectric minerals with polar axes (that is, axes whose two ends are dissimilar) will, under pressure, acquire an electric charge, positive at one end, negative at the other. This phenomenon is known as 'piezo-electricity'. If tension is applied along the polar axis, the charge is reversed.

Quartz, being highly piezo-electric, is used in apparatus such as oscillators and impulse-generators, and in the making of quartz watches.

Certain crystals, subjected to variations of temperature, produce electric charges on some of their surfaces. This phenomenon is known as 'pyro-electricity', and a typical pyro-electric mineral is tourmaline. The charges may be observed by means of an electrometer, or, more conveniently, with a dust of red lead and sulphur blown vigorously through a screen of thin silk to charge the particles with static electricity. The red lead, having received a positive charge, will then settle on the negatively-charged part of the mineral, while the negatively-charged sulphur goes to the part possessing a positive charge.

MAGNETISM

Magnetic minerals are those which respond to the action of a magnet. Minerals with a strong response are ferromagnetic, those with a weaker response paramagnetic; diamagnetic minerals are actually repelled by a magnetic field.

Magnetite is ferromagnetic. Haematite and siderite are examples of paramagnetic minerals, while sulphur, titanite, quartz and many others are diamagnetic.

The innate properties of magnetite make it a long-lasting natural magnet. In a polar specimen the magnetism at the extremities is opposite; when the magnetism is identical, the specimen is bipolar.

OPTICAL PROPERTIES

Incident light falling upon any object is partly absorbed and partly reflected back, and so the lustre of crystals will vary with the intensity of reflection. Their grading according to degrees of lustre, as follows, will help with the identification of minerals:

adamantine	—very brilliant, as in diamonds
subadamantine	—slightly less brilliant, as in linarite
vitreous	—a common degree of brightness, as for quartz
metallic	—when, as in metals, much of the light is absorbed
submetallic	—minerals of this lustre are opaque, but thin and transparent in section, as is cinnabar
pearly	—slightly irridescent, as in the micas
silky	—typical of fibrous minerals such as asbestos
resinous	—sparkling and tending to yellow, as sulphur
greasy	—subdued, as in massive talc and nepheline
dull	—giving off very little lustre, e.g. chiolite. Samples with this lustre are nearly invisible in water

COLOUR

This is a useful indication for idiochromatic minerals only, whose own colour is unvarying. Most minerals, however, are allochromatic; that is, their colour varies according to which atoms are present in, or missing from, the crystal lattice. Idiochromatic minerals, such as sulphur, cuprite, malachite, cinnabar and dioptase, keep their colour, slightly softened, in powder form. Allochromatic minerals in powder form are usually grey, or whitish.

LUMINESCENCE

Some minerals, when energized by mechanical or chemical means, or by heat, emit certain wavelengths and assume colourations. This is the phenomenon of luminescence. Phosphorescence occurs if the emission continues after the exciting energy is withdrawn, fluorescence if it ceases with the stimulus. The spectacular fluorescence known as Wood's Effect is the result of short-wave radiation on certain radioactive minerals.

Fluorescence can be induced with portable ultraviolet lamps and is important in the search for radioactive minerals for industrial use.

Minerals such as autunite and scheelite are permanently fluorescent. Others—scapolite, calcite and fluorite among them—are so only when they contain impure particles, known as activators, or when structural deficiencies are present.

REFRACTION AND DOUBLE REFRACTION

Light penetrating a transparent crystal changes speed, and its ray approaches the normal of the bounding surface. This is the phenomenon we call refraction.

Monometric, or cubic, crystals are monorefractive, but dimetric and trimetric crystals behave differently. Light penetrating them divides itself into two rays: the ordinary ray, which obeys the laws of optics, and the extraordinary one: the light has been resolved into two plane-polarized rays. Ordinary light vibrates in indefinite planes along the line of propagation; polarized light vibrates in one plane only. The plane at right angles to that in which light vibrates is called the plane of polarisation. The two rays, ordinary and extraordinary, vibrate in planes at right angles one to the other, and at differing speeds.

The optical analysis of crystals depends to a great extent on the establishment of refractive indices (see Further Reading for textbooks on mineralogy) and upon microscope observations in polarised light. Polarised light may be produced with tourmaline tongs (*Fig* 22), with nicols or with artificial polaroids.

Fig 22 Tourmaline tongs

The tongs consist of two thin sections of tourmaline, cut parallel to the optic axis, and inserted into revolving discs to make a relatively simple laboratory instrument. The first disc receives light and polarizes it, splitting it into two rays, one of which is absorbed while the other passes on. If the analyser-section through which we look is held parallel to the other, light is seen; when it is held at right angles, with both sections lying along the perpendicular optic axis, light is occluded.

The same result is achieved with nicols, which are obtained from Iceland spar, a very clear calcite. When the pieces of nicol are parallel there is light; when they are crossed, there is none.

If a sliver of crystal is introduced between the nicols when these are crossed and the light is parallel, we may see either darkness or light: darkness if the crystal is from a non-birefractive mineral of the cubic system, light if it belongs to the dimetric or trimetric groups, which are birefractive. In the latter case, light from the polarising nicol is separated into ordinary and extraordinary rays and these interfere with the working of the analyser nicol which gives light. To discover whether our specimen is dimetric or trimetric we insert an Amici lens between the crossed nicols. The lens will cause the light to converge and allow us to see the interference figures characteristic of uniaxial or biaxial minerals (*Fig* 23). The phenomenon is evident when the specimen is cut normally (ie at 90°) to the optic axis in uniaxial crystals and, in biaxial, perpendicularly to the bisectrix of the two optic axes.

Optic axes are the directions in a crystal along which birefraction does not occur. A diametric crystal has one such axis, a trimetric crystal has two. The optical observation of crystals under the

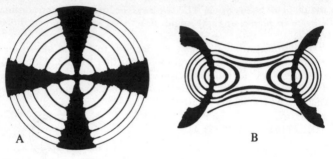

Fig 23 A = uniaxial birefractive crystal
         B = biaxial birefractive crystal

mineralogical microscope is immensely informative, helping us, for instance, to identify the separate mineral components in tiny slivers of rock.

RADIOACTIVITY

Many minerals contain radioactive elements such as uranium, thorium and radium, and are in a continuous state of transmutation, emitting alpha and beta particles and gamma rays. They will affect a photographic plate and register on a Geiger counter. Autunite, torbenite, carnotite, betafite and uranite are among the many radioactive minerals.

3. CHEMICAL TESTS

Qualitative and quantitative chemical analysis of specimens may take considerable time and effort, but shorter and valuable experiments are possible with the aid of a few reagents and a few inexpensive items of equipment. All we need are some test-tubes, a small loop of platinum wire fused onto a handle, a blowpipe, Bunsen burner, spatula, penknife and an agate mortar.

Necessary chemicals include hydrochloric acid, nitric acid, sodium carbonate, cobalt nitrate, potassium iodide, hydrogen peroxide, sulphur, lead dioxide, barium chloride and ammonia.

We may begin by putting a little powdered mineral in a test-tube with some drops of hydrochloric acid; immediate effervescence will tell us the mineral is a carbonate. If the cold powder gives no result we can heat it, because some minerals, such as dolomite, effervesce only when hot.

To test for true native gold, a black flint is rubbed first with an attested piece of gold, then with the specimen to be examined, and a few drops of nitric acid are poured along the lines so made. Since gold is unaffected by nitric acid, both speciments are gold if both lines remain visible; if the line traced by the doubtful specimen is destroyed, that specimen is not true gold.

On the following two pages are brief but necessary notes on some of the many chemical tests we can apply to minerals.

IDENTIFICATION OF SULPHIDES

Physical data are helpful here: sulphides are soft and opaque, with metallic lustre and high specific gravity. They will burn in air, producing sulphur dioxide, SO_2, with its characteristic smell. When they are heated in a closed tube, flowers of sulphur are sublimated on its sides. Sulphides are soluble in acid and the addition of hydrochloric acid gives hydrogen sulphide, H_2S, evil-smelling and poisonous.

IDENTIFICATION OF HALIDES

If these are chlorides, and soluble in water, a solution of silver nitrate, $AgNO_3$, will produce a white precipitate of silver chloride, $AgCl$.

If insoluble in water, we test with nitric acid, HNO_3, and silver nitrate, $AgNO_3$. White flakes of silver chloride, $AgCl$, are formed on heating.

When testing fluorides, add sulphuric acid, H_2SO_4, to a little of the powdered mineral. With slight heating, white fumes of hydrogen fluoride, HF, are emitted and will etch the glass tube. Care should be taken not to inhale these acidic fumes.

The test for iodides is to add potassium hydrogen sulphate, $KHSO_4$, to the powdered mineral, which will then give off a violet-coloured iodine vapour.

IDENTIFICATION OF CARBONATES

These will produce carbon dioxide (CO_2) on the addition of acids in conditions of cold or heat. If brought to great heat they will calcify into their respective oxides and produce CO_2.

Calcite and aragonite, though chemically the same, crystallize in different ways. If this different habit is not identifiable because both samples are microcrystalline (ie with crystals visible under the microscope), the two minerals can be distinguished by means of the Meigen test. For this test a solution of cobalt nitrate, $Co(NO_3)_2$, is added to a little powdered mineral in the test-tube and held for several seconds at boiling-point. Aragonite powder will turn violet in colour, while calcite remains white.

IDENTIFICATION OF BORATES

When these are treated in powdered form with sulphuric acid they colour a flame green.

IDENTIFICATION OF SULPHATES

Those soluble in water or hydrochloric acid will produce the white precipitate, barium sulphate, $BaSO_4$, when added to a solution of barium chloride, $BaCl_2$.

The 'hepar sulphuris' test, used also on sulphides, is applicable to non-soluble sulphates. The powdered mineral is mixed with sodium carbonate, Na_2CO_3, and heated strongly over charcoal. The residue, placed on a sheet of silver and washed in water, will produce hydrogen sulphide, H_2S, which causes the silver to turn black. (This residue also contains sodium sulphide, Na_2S.)

IDENTIFICATION OF PHOSPHATES

These will melt in the flame of a blowpipe, tingeing the flame green. They are soluble in nitric acid, HNO_3, and if ammonium molybdate is added to the solution the result is a yellow precipitate of ammonium phosphomolybdate.

IDENTIFICATION OF ARSENATES

These melt in the flame of a blowpipe; the vapour produced will smell of arsenic and the flames assume a bluish tinge.

To distinguish between phosphates and arsenates, we add magnesium acetate, $(CH_3COO)_2Mg$, to the ammoniacal solution of the mineral, strain the precipitate and add silver nitrate, $AgNO_3$. A phosphate will then turn yellow, an arsenate reddish-brown.

IDENTIFICATION OF VANADATES

These will dissolve in dilute nitric acid, HNO_3.

The solution, evaporated on a thin metal sheet, leaves a red deposit of divanadium pentoxide, V_2O_5. If hydrogen peroxide is added, the solution becomes reddish-brown.

FLAME TESTS

Rapid and useful tests over flame will indicate what elements are present in a given mineral.

The atoms excited by the heat of a flame contain, on their outermost orbital zones, electrons in movement. These, as they recede to an inner zone, emit absorbed energy in the form of light-rays. To make our test, we dip the platinum wire in hydrochloric acid, take up a pinch of powdered mineral on the wire and put it in the flame:

Sodium will give a steady yellow flame
Calcium will give an orange-red flame
Copper will give a bright green flame
Barium will give a steady yellow-green flame
Strontium will give a bright red, spurting flame
Lithium will give a vermilion-red flame
Potassium will give a violet flame

A fragment of chalcopyrite can, for instance, be distinguished from pyrites by the copper-green of the flame in a quick test of this sort.

MINERAL CLASSIFICATION

Some 2000 mineral species are at present known and 20 or so new species are discovered annually all over the world. Recent finds in Italy alone, for instance, include roggianite, canavasite, liottite, honoratoite and merlinoite. Since each species may contain several varieties, arising usually from the presence in minute quantity of an element not normally part of its crystalline structure, the total number of separate minerals is obviously very great.

Mineralogists classify this wealth of species and varieties in different ways, according to genetic, physical, chemical or structural characteristics, or to the data of crystallochemistry; while minerals may be named, surprisingly at times, in someone's honour, or after their place of discovery, or after the discoverer himself.

One of the most generally adopted systems is that of the German Hugo Strunz (*Mineralogische Tabellen*; 5th ed., Leipzig, 1970). This is a crystallochemical classification based on structural type and chemical composition. Minerals are divided into 9 classes, with each class subdivided into groups, each group into series and each series into families, including specific single units. The 9 mineral classes are:

1. Elements (together with compounds, carbides, nitrides, phosphides)	50 species approx
2. Sulphides (with selenides, tellurides, arsenides, antimonides and bismuthides)	300 species approx
3. Halides	100 species approx
4. Oxides and hydroxides	250 species approx
5. Nitrates, carbonates, borates	200 species approx
6. Sulphates (with chromates, molybdates, wolframates)	200 species approx
7. Phosphates, arsenates, vanadates	350 species approx
8. Silicates	500 species approx
9. Organic substances	20 species approx

The most widely distributed minerals are the silicates, which make up the greater part of the lithosphere, and the 6 subdivisions of this class are based not on chemical composition but on silicate structure.

All silicates contain silicon, Si, which is tetravalent, having ionic bonding to four oxygen atoms.

From structural analysis we learn that the silicon ion, Si^{4+}, is always at the centre of a tetrahedron whose vertices are tenanted by four ions of oxygen, O^{2-}. The basic silicate structure, $(SiO_4)^{4-}$, is formed in this way. The subdivisions of this class are:

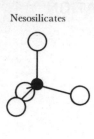

Nesosilicates

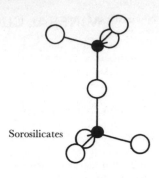

Sorosilicates

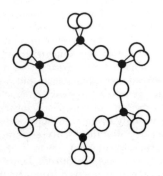

Cyclosilicates

Amphibolic inosilicates

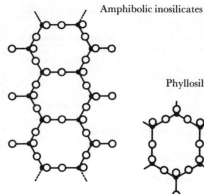

Phyllosilicates

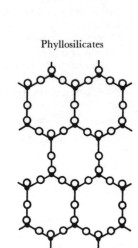

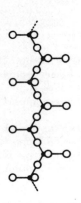

Pyroxenic inosilicates

1. **Nesosilicates** Here the tetrahedrons are isolated and distinct, and every free valency is saturated with cations. Examples: olivine, the garnet group, topaz, zircon.
2. **Sorosilicates** In these the tetrahedrons are grouped, each pair being linked by one oxygen and thus creating the structure $(Si_2O_7)^{6-}$, with the six free valencies forming a chain. Examples: hemimorphite, epidote, vesuvianite, ilvaite.
3. **Cyclosilicates** Rings are formed of 3, 4 or, more usually, 6 tetrahedrons. Examples: beryl, axinite, tourmaline, dioptase.
4. **Inosilicates** Also known as fibrous silicates. Here the tetrahedrons form chains. In the pyroxene group the chains are single, or infinite; in the amphibole group they are double, or finite. The minerals have a characteristic fibrous appearance. Examples: hedenbergite, diopside (pyroxenes); pargasite, hornblende (amphiboles).
5. **Phyllosilicates** The three base-points of each tetrahedron, around the silicon atom, are attached to the base-points of neighbouring tetrahedrons, so that sheets or layers are formed. The formula $(Si_2O_5)^{2-}$ indicates a structure in which the ratio of silicon to oxygen in each unit is 2 to 5. Examples: talc, chrysotile, kaolinite, the mica group.
6. **Tectosilicates** In these the tetrahedrons join at all four corners. Whereas in layer silicates the tetrahedrons are arranged three-dimensionally, the arrangement here would leave no free valency to the oxygen atoms—which could not, therefore, attach cations—were it not for the fact that the tetravalent silicon atom is often replaced by one of trivalent aluminium.

The chemical formula for orthoclase, $K(AlSi_3O_8)$, shows, for instance, that the atom Al breaks the equilibrium and sets free an oxygen valency, which can then attach the cation K. Examples: orthoclase, albite, anorthite, leucite, analcime.

Tectosilicates

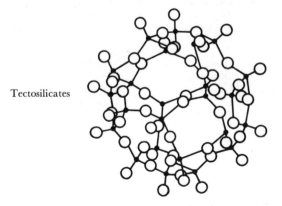

PREPARATION AND CARE OF MINERALS

Those who acquire shining-bright specimens which do not need complicated cleaning are fortunate indeed. Often a mineral is covered in clay, silica or organic substances of one sort or another; every specimen has its own natural setting and under different conditions may, sooner or later, lose stability, and change. While diamond, quartz, casserite and gold are very stable, and most minerals relatively so, there are some which lose their original brightness in a cabinet and crumble away. Native iron will rust on exposure to the air; pyrites and marcasite oxidize and turn to powder.

Sulphides, which are formed in the absence of oxygen, are most seriously endangered by air; oxides, on the other hand, are among the most stable of minerals.

Over the years, many museums and collectors have found various methods of dealing with these problems. Here we shall mention only the principal, and simplest, ways of cleaning mineral specimens and of protecting them from deterioration.

1. DUST

This is the main enemy of samples in any collection: if the minerals have to be dusted too often, they run the risk of damage through mishandling, quite apart from the fact that it is almost impossible to dust certain fibrous and delicate species like cuprite, malachite, mesolite, whether one blows air on them or tries to remove the dirt with a vacuum cleaner. There are many species thus structured; among them are:

Alotrichite	Goslarite	Natrolite
Auricalcite	Gowerite	Scolecite
Boothite	Gypsum	Strunzite
Boulangerite	Hydrozincite	Thomsonite
Bronchantite	Jamesonite	Ulexite
Cyanotrichite	Melanterite	Uranophane
Epsomite	Meneghinite	Zeolites
Erionite	Millerite	Zinckenite
Gonnardite	Mordenite	

It is therefore necessary to protect these samples by keeping them in transparent boxes.

If the sample is rather tougher, it is usually brushed with a soft camel-hair brush best suited to remove the dust. Sometimes dirt has to be removed by scrubbing, in which case harder brushes, even metal ones, should be used; never use nylon or plastic brushes, as these leave traces, particularly in conjunction with solvents. Good results are achieved by using toothpaste, pumice powder or other cleansing powders. Soap or detergent can also be used if this does not harm the sample.

Many porous or clayish minerals must be simply brushed, without wetting, as they are water-absorbent and could disintegrate. The following are either porous or clayish*:

Allophane	Roemerite	Taylorite
Ammonium salt	Rossite	Thenardite
Beidellite*	Saponite*	Thermonatrite
Caolinite*	Sassolite	Tincalconite
Dickite*	Sauconite*	Trona
Halloysite*	Schairerite	Tschermigite
Hectorite*	Schroeckingerite	Tychite
Hisingerite*	Searlesite	Ulexite
Illite*	Senarmontite	Valentinite
Montmorillonite*	Sideronatrite	Vanthoffite
Nacrite*	Silvite	Villiaumite
Nontronite*	Sodium alum	Voltaite
Pintadoite	Sodium nitre	Yukonite
Pisanite	Sulphoborite	Zinkosite
Polialite	Syngenite	Zirklerite
Potassium alum	Szomolnokite	
Rinneite	Tachyhidrite	

A great number of minerals are soluble in water and would be ruined by washing; the following varieties would be thus affected:

Alunogene	Cotunnite	Hieratite
Afthitalite	Criolite	Hydrophilite
Amarantite	Cupromapnesite	Ilesite
Ammonium alum	Darapskite	Inyoite
Apjohnite	Dietrichite	Jarosite
Arsenolite	Dietzeite	Kainite
Arsenosiderite	Dolerophanite	Kaliborite
Autunite	Douglasite	Kalinite
Bilinite	Epsomite	Kernite
Bioedite	Ettringite	Kieserite
Bischofite	Fernadinaite	Koenenite
Borax	Gaylussite	Kornelite
Botryogen	Gerhardtite	Krausite
Boussingaultite	Glauberite	Kyanocroite
Burkeite	Goslarite	Lantanite
Carnallite	Greenockite	Larnite
Chalcokyanite	Gypsum	Lautarite
Chalk	Halite	Lecontite
Colemanite	Halotrichite	Leonite
Copiapite	Hanksite	Leowite
Coquimbite	Hesaidrite	Mallardite

Melanterite	Oxammite	Sodium nitre
Mendozite	Palmierite	Sulphurborite
Metahewettite	Pascoite	Syngenite
Metarossite	Phillipite	Szomolnokite
Metavoltine	Pickeringite	Tachyidrite
Minasragrite	Picromerite	Taylorite
Mirabilite	Pintadoite	Teschemacherite
Misenite	Pisanite	Thenardite
Mitscherlichite	Polialite	Thermonatrite
Morenosite	Potash alum	Tincalconite
Nahcolite	Rinneite	Trona
Natrocalcite	Romerite	Tychite
Natron	Rossite	Ulexite
Nitre	Saxolite	Vanthoffite
Nitrobarite	Schroeckingerite	Villiaumite
Nitrocalcite	Siderotile	Voltaite
Nitromagnesite	Silvite	Winebergite
Orpiment	Sodium alum	Zirklerite

Mica, although not soluble, should never be washed as it would separate into laminae; neither should the sulphurs be washed, as they would oxidize: they should be brushed with a dry, soft, camel-hair brush.

Capillary crystals, as long as they are not soluble in water, can be cleaned by immersing them into demineralised water, rinsing them slowly and drying them in the sun or air. Soaps or detergents dissolved in water are always better than any acid, soap being by far the best for gold. Better results can be obtained by using lukewarm water, as the combination of detergent and hot water can produce oxygen which may in turn cause discoloration.

Even if a mineral is only partially soluble in water, it should be cleaned with such detergents as benzene, kerosene or carbon tetrachloride, but one should take care not to inhale their vapours and remember that they are highly inflammable. *Tourmaline* can be cleaned of clay with kerosene followed by a quick wash in water.

Quartz geodes should be placed in a warm solution of ammonia and distilled water, which will give them a sheen without leaving haloes. The iridescent geodes of *brown calcite* can only be cleaned with brushes and by blowing air onto them, as any liquid would destroy their iridescence. *Galenas* can easily be cleaned with ammonium acetate and a good rinse in running water or silver-cleaning products can be used. *Dioptase* should be washed in acetic acid; hydrochloric acid would damage it.

If the minerals are not too fragile, ultrasonic generators can be used with good results; the length of their application varies between one and five minutes.

Tar can be removed with caustic soda applied for a length of time varying between one and thirty minutes. Once the sample has been cleaned it should be immersed in detergent and brushed with a soft brush.

2. RUST

Many of the yellow, brown or red hues which deface some samples are iron oxides, or rusts. Minerals can be protected from them by covering their surface with lacquer. Bearing in mind that no acid should be used for calcium carbonate (calcite, aragonite, marble, etc), rusts can be removed mainly with sodium citrate, sodium hydrosulphate, oxalic acid, hydrogen sulphide and tartaric acid. Individual stains can be removed with a solution of sodium citrate (one part in six of water): cotton wool impregnated with it should be kept on the stains for a quarter of an hour. Hydrogen sulphide should be used in the same way.

Pink chalcedony can be cleaned with detergents and sodium hypochlorite. A solution of these, warmed and then cooled again, should be applied and rubbed with oxalic acid. This particular acid can be used in several ways, one of the simplest being as follows: rinse and brush the quartzes in running water, dip them into vinegar and then into ammonia and rinse them well in water. Alternate this treatment with prolonged immersions into oxalic acid, and finally dry thoroughly. The quartz can also be immersed in a 50 per cent solution of hydrochloric acid for an hour, then left overnight in oxalic acid.

Hydrogen sulphide is the best means of eliminating iron oxides, but it is poisonous, and the procedure too complex to be dealt with here. Some collectors recommend the use of tartaric acid, rather than oxalic acid, in the cleaning of *quartz*; it is indeed stronger, and does not produce any discoloration or insoluble oxalates.

3. CLAY

Clay crusts on the surface of mineral samples can be removed with hard brushes, as long as the minerals are not calcite, gypsum, selenite or other porous species. Ordinary vinegar can often remove clay; if this fails, one could try a bleaching solution. One way of doing this is to boil the sample in a concentrated solution of sodium sulphide, and, once cool, wash it in lukewarm water.

4. SILICA

Silica contained in water can form a veil on quartzes and other minerals. If the affected mineral is not likely to be damaged by acids or water, the sample can be treated with a solution of hydrofluoric acid in a plastic container; once the silica has been removed, the sample should be rinsed in water and left in distilled water overnight. Care should be taken when handling hydrofluoric acid as it is dangerously corrosive.

5. ORGANIC SUBSTANCES

On the whole these are represented by encrustations of lichens. The sample thus affected should be brushed with aqueous ammonia; should the lichens resist this, one can use concentrated sulphuric acid—as long as the mineral is not soluble in acids or water. Sulphuric acid should be left to work for one hour, then rinsed off, and the sample brushed and left in distilled water for a week to eliminate all remaining traces of organic matter.

6. CLOUDING

This is due to various chemical reactions occurring on the surface of certain metals, mainly copper and silver. The blackening of *copper* is caused by oxidation; a green patina indicates copper sulphates, hydroxides and carbonates. One method of cleaning copper involves using a solution of one part acetic acid to ten parts water. The sample should then be immersed in the solution, rinsed in distilled water and brushed with a hard brush if the crystals are close together. Hydrochloric and sulphuric acids could damage the mineral.

The British Museum recommends the following method:

1. Mix one part by weight of caustic soda with three parts double tartrate of sodium and potassium.
2. Dilute this in 20 parts distilled water.
3. Suspend the sample in the solution with a copper thread tied round it, and move it about occasionally. (This will take from half an hour to one hour.)
4. Rinse the sample in running water, then leave in distilled water for one hour.
5. Brush the sample (if this is not dangerous), rinse it, and leave to dry in the air.

Silver is usually covered by a patina of silver sulphide and can be cleaned using the following method. This also applies to copper and is practised by jewellers:

Dissolve $\frac{1}{4}$ teaspoon of sodium cyanide in half a litre of distilled water. Cyanide is a powerful poison and must not be touched. Care must also be taken not to inhale its vapour and it must not come into contact with acids. The sample should be left in the solution overnight, then washed in soapy water, rinsed in running water and dried in the air. If the sample is unlikely to be damaged, it may be brushed, washed in soapy water and rinsed thoroughly. Once cleaned, silver and copper samples should be covered with protective lacquers.

7. OXIDATION

Air and humidity cause the oxidation of iron sulphides, which turn to sulphates. Collectors can be greatly troubled by the oxidation of pyrites and marcasites: if the samples are fractured, the degenerating process is faster and often leads to nearby samples being damaged. The oxidation liberates sulphuric acid which attacks labels, supports and even the drawers themselves. Certain localities produce only stable or only unstable pyrites (for example, unstable pyrites only are found on the island of Elba). Not all samples deteriorate rapidly.

Pyrites should be protected with a good layer of lacquer, repeated from time to time. Should the sample need cleaning prior to lacquering, it can be immersed for 10–15 minutes in pure hydrochloric acid within a closed container. The process should be repeated with fresh acid until the sample no longer reacts; it should then be dried and immersed twice in ether. It should then be lacquered or immersed in a plastic solution of acetate and toluene with 7 per cent by weight of vinyl acetate.

Another method is to brush the sample with a hard brush dampened with oxalic acid (4 teaspoons in a $\frac{1}{4}$ litre of water). Once clean, the sample should be rinsed in boiling water: the heat absorbed by the pyrite will help it to dry quickly. It should then be lacquered.

Marcasite, not nearly so common as pyrite, oxidizes in the same way. Again, certain localities are known to produce only stable or only unstable marcasite. This mineral should be cleaned with a dry brush to avoid oxidation, and using the same methods recommended for pyrite. Once lacquered, both minerals should be kept in closely sealed boxes to avoid contact with air; transparent plastic boxes make ideal containers. *Meteoric iron* also requires the same care as pyrite and marcasite, as it oxidizes easily. Other minerals which oxidize and should be lacquered are alabandite, bravoite, cloantite, gersdorffite, polydymite, rammelsbergite and smaltite.

8. DELIQUESCENCE

Deliquescent minerals are those which, once in contact with air, attract and absorb humidity and finally liquefy—a serious danger for collectors. They should be treated with dehydrating substances (quicklime, chloride of lime), sealed within glass containers or tightly fitting plastic boxes. The deliquescent minerals are:

Bischofite	Huantajayte	Nitrobarite
Carnallite	Hydrophilite	Nitrocalcite
Chlorocalcite	Kainite	Nitroglauberite
Chloromagnesite	Kieserite	Nitromagnesite
Chloromanganokalite	Kremersite	Rinneite
Darapskite	Lawrencite	Scacchite
Douglasite	Melanovanadite	Silvite
Eritrosiderite	Minasragrite	Soda-nitre
Gerhardtite	Molysite	Thermonatrite
Halite	Nesquehonite	
Hisingerite	Nitre	

Several mineral species turn into different ones under the action of air and humidity. Thus arsenic becomes arsenolite, niccolite becomes annabergite, and so on.

9. EFFLORESCENCE

Minerals which lose their water content and crumble when exposed to the air are described as efflorescent. They change their physio-chemical characteristics and often become different varieties.

Melanterite becomes siderolite, morenosite turns into retgersite, and so on. These efflorescent minerals include:

Bianchite	Hinyoite	Phosphorroesslerite
Boothite	Hydromagnesite	Pirssonite
Boracite	Kernite	Pisanite
Borax	Lansfordite	Probertite
Calcantite	Lantanite	Rhomboclase
Chalcocyanite	Laumontite	Spurrite
Coquimbite	Mallardite	Struvite
Epsomite	Melanterite	Szmikite
Gaylussite	Minasragrite	Trona
Goslarite	Mirabilite	Tschermigite
Halite	Morenosite	Voltaite
Halotrichite	Natron	

It is worth noting, however, that these varieties are not always efflorescent. In order to preserve them properly, the samples should be placed in sealed plastic boxes; efflorescence can be prevented by placing some cotton wool soaked in water, inside the box, or by lacquering the sample before storing it.

10. LIGHT

Minerals which can be damaged by light are described as photosensitive. Most minerals, in the long run, suffer from the effects of light, but some are affected even by brief exposure. Several minerals are seriously damaged by ultraviolet rays and radioactivity. The minerals which can be damaged by light and air (oxygen) are:

Acantite	Eritrite	Polibasite
Aguilarite	Fizelyite	Polidimite
Alabandite	Freieslebenite	Pirargirite
Alaskaite	Graftonite	Pirostilpnite
Anapaite	Hauntayaite	Proustite
Andorite	Hessite	Ramdorite
Aramayoite	Hureaulite	Ratite
Argentite	Hutchinsonite	Realgar
Argyrodite	Kleinite	Samsonite
Baumhauerite	Koninckite	Sartorite
Berzelianite	Legenbachite	Silvanite
Bromargirite	Lorandite	Simplesite
Canfieldite	Marshite	Smithtite
Cerargirite	Matildite	Stephanite
Chalcocite	Miargyrite	Stibnite
Cinnabar	Miersite	Stromyerite
Crocoite	Montroydite	Terlinguaite
Cuprite	Nantokite	Trechmannite
Diaforite	Naumannite	Vivianite
Dietzeite	Pearceite	Vrbaite
Dufreynosite	Penroseite	Xantoconite
Eglestonite	Phoenicocroite	

All these minerals should be kept in dark drawers or in a maximum of shade.

Most minerals, in the long run, change colour under the influence of light; calcite, quartzes, turquoise, topaz and fluorite all change colour to a lesser or greater degree according to their original locality.

11. CRACKING AND SCRATCHING

Minerals likely to crack should be preserved in glycerine. Crystals which have been scratched should be well brushed and then lacquered.

12. LACQUERING

A good lacquering keeps minerals free from rusts, oxidation, humidity and dust.

The sample should first be carefully brushed. There are many transparent lacquers and synthetic resins which can be applied to minerals. Lacquers tend to evaporate and should be applied in two thin layers (with a 24-hour interval between each layer); the sample should be well protected from dust until the lacquer has dried out completely.

Other collectors prefer to use a solution of 10 per cent vinyl acetate in equal parts of toluene and acetone. Samples consisting of groups of crystals should be hung on a thread and dipped in the solution for a few minutes only. Immersion is absolutely necessary for porous minerals, which should not be touched for at least 24 hours and allowed to dry in a warm environment. If the solution is applied with an atomizer, it should be further diluted.

Nowadays one can obtain commercially excellent cellulose acetate solution already contained in atomizers and therefore ready to use. Acetone can be used to remove the varnish. Some people use seven parts xylol to one Canada balsam; this solution is particularly suited to *marcasite*, as long as it is well cleaned and dried . It can also be applied with a brush in two or three layers, allowing a few hours to elapse between applications. 'Brasolin' is an excellent commercial protective varnish, which can be applied by spraying and does not contain benzene.

HOW TO READ THE SYMBOLS

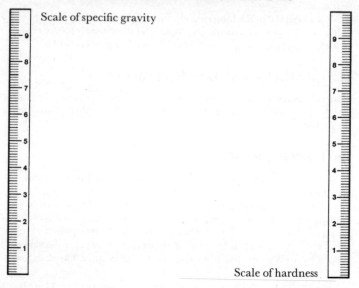

Scale of specific gravity

Scale of hardness

MINERAL CLASSES

1	**I**	Native elements
2	**II**	Sulphides
3	**III**	Halogenures and Halides
4	**IV**	Oxides and Hydroxides
5	**V**	Carbonates, Nitrates, Borates
6	**VI**	Sulphates, Chromates, Molybdates, Wolframates
7	**VII**	Phosphates, Arsenates, Vanadates
8	**VIII**	Silicates

EXCITATION AND RADIOACTIVITY

9  Flourescent minerals

10 ☢ Radioactive minerals

CRYSTAL SYSTEMS

11   Cubic

12 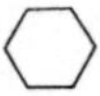   Hexagonal

13   Trigonal

14   Tetragonal

15    Rhombic

16 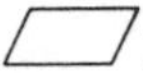   Monoclinic

17    Triclinic

CLEAVAGE

18    Perfect

19    Imperfect

20    Non-existent

COLOUR OF LIGHT-REFLECTING MINERALS

21   White

22   Grey (whether metallic or not)

23   Black

24   Yellow

25   Green

26   Rusty brown

27   Blue

28   Red

29   Orange, pink

30   Violet, purple

COLOUR OF THE POWDER

31 ⬔ White or pale

32 ⬔ Grey

33 ⬔ Black

34 ⬔ Yellow

35 ⬔ Green

36 ⬔ Rusty brown

37 ⬔ Blue

38 ⬔ Red

39 ⬔ Orange, pink

40 ⬔ Violet, purple

LUSTRE

41 ☀ Adamantine

42 ☀ Subadamantine

43 ☀ Vitreous

44 ☀ Metallic

45 ☀ Submetallic

46 ☀ Pearly

47 ☀ Silky

48 ☀ Resinous

49 ☀ Greasy, waxy, oily

50 ☀ Dull

FUSIBILITY

51 ⌇ Fusible

52 ⌇ Resistant to fusion

53 ⌇ Non-fusible

COLOUR IN A FLAME

54 Yellow

55 Green

56 Red

57 Orange

58 Violet

59 Yellow-orange

60 Yellow-green

SOLUBILITY

61 Soluble in water

62 HCl Soluble in hydrochloric acid

63 HNO₃ Soluble in nitric acid

64 H₂SO₄ Soluble in sulphuric acid

65 HF Soluble in hydrofluoric acid

66 Insoluble in acids

OCCURRENCE

67 Very rare

68 Rare

69 Common

70 Very common

ORIGIN

71 In igneous rocks

72 In metamorphic rocks

73 In sedimentary rocks and hydrothermal deposits

USE

74   Metallurgy, building industry, etc

75   Jewellery

76   Chemical industries

77   Private collectors, scientific research

POLARISED LIGHT WITH CONVERGING NICOLS

78   Interference figure of dimetric crystals

79   Interference figure of trimetric crystals

PRESERVATION

80   **!**   Perishable

PLATES

The minerals have been grouped according to their dominant colour which is indicated in the top corner of each page for ease of identification. It must be pointed out, however, that some specimens can occur in different colours. The samples illustrated are housed in the Museo di Storia Naturale Don Bosco, Turin, of which the author is Director. The photographs are from his collection.

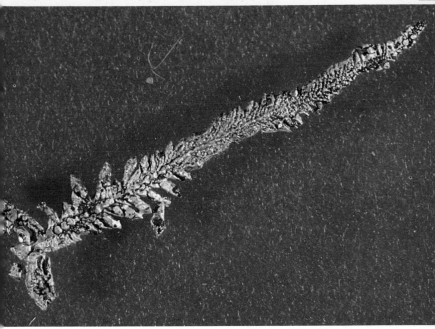

Ag   Rjoia (Argentina)

Silver. High specific gravity: 10.5. Ductile, malleable, bright metallic lustre. Fusible at 960°C. Occurs in filaments, dendrites, rarely in crystals. The samples from Kongsberg (Norway) are quite unusual; it can be found at Freiberg and Schneeberg (E. Germany), Andreasberg and Wittichen (W. Germany); as well as at Pribram (Czechoslovakia) and Joachimsthal (E. Germany). Mexico heads world production with deposits at Chihuahua, Guanajuato and Batopilas. Beautiful samples have been found in Copiapo and Canarcillo (Chile) and at Rjoia (Argentina). At Cobalt (Canada) a block weighing 744kg was found. Silver mines are active in the USA (Silver King—Arizona), Australia (Broken Hill), Siberia, and Italy (small quantities in Sarrabus, Sardinia).

10,5

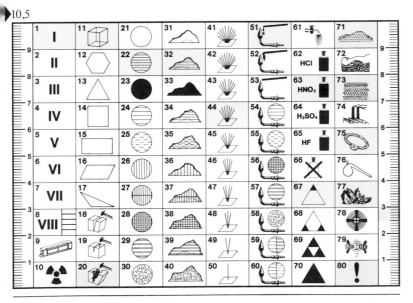

Fe, Ni   Canõn Diablo, Arizona (fell in 1876)

Iron-nickel. Iron is either native to the earth's crust or of meteoric origin. The most famous source of native iron is Disko (Greenland) where magmatic masses have enclosed coal banks and transformed iron oxides already present in the rocks. It can also be found, although in lesser quantities, in Mühlhausen (E. Germany), Bühl (W. Germany), Rouno (Poland), on the island of St Joseph (Lake Huron, Ontario) and in the coal deposits of Missouri. If meteoric iron, which is rich in nickel, is treated with acids on a section, it reveals the Widmannstätten figures: in other words, nickel comes to the fore, being less liable to attack than iron.

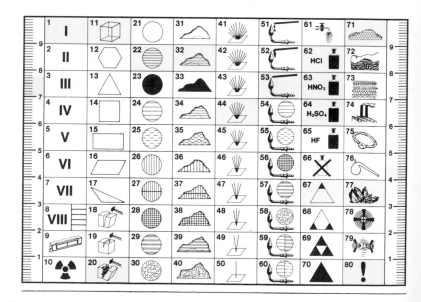

Ag_2S Czechoslovakia

Silver sulphide. Malleable, soft, can easily be cut with a pen-knife. Can be found as distinct crystals, aggregates, and dendritic forms. Important deposits are at Kongsberg (Norway), Schemnitz, Kremnitz and Pribram (Czechoslovakia), Joachimsthal and Freiberg (E. Germany) and Andreasberg (W. Germany). A certain amount is found at Colquechaca (Bolivia), Chanarcillo and Atacama (Chile), various localities in Peru and Mexico (Pachuca, Zacatecas, Guanajuato), at Aspen and Leadville (Colorado) and Sarrabus (Sardinia).

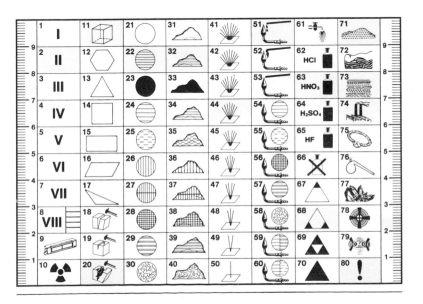

ZnS   Rumania

Zinc sulphide. Also called blende. Often contains cadmium, iron (= marmatite). There are several deposits in Europe and elsewhere. Splendid crystals have been found in the mines of Trepca (Yugoslavia), Kapnik (Hungary), Pribram (Czechoslovakia), Bleiberg (Austria), Santander (Spain) and Alston Moor (England). The richest deposits in the world are in Kansas, Oklahoma, Missouri (USA), Broken Hill (NSW, Australia) and Sullivan (Canada). Fine deposits exist in Bolivia, Peru, Mexico, Italy (Rosas, good crystals; Iglesiente, Sarrabus, Serravezza, Brembana and Seriana valleys).

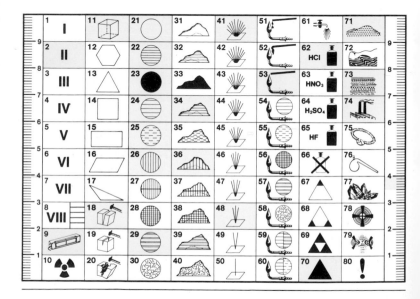

$Cu_3AsS_{3.25}$ Mexico

Copper arsenic sulphide. Produces isomorphic mixtures with tetrahedrite. These two minerals are called grey copper because of their colour. Often contains silver (= binnite), bismuth, mercury, cobalt and nickel. Can be a good source of copper. Grey copper can be found at Cabriere (France), Freiberg (E. Germany), Kapnik (Hungary), Cornwall (England), Modum and Skutterud (Norway), Boliden (Sweden) and Swatz (Austria). It can also be found at Tsumeb (Namibia) and various localities in the USA, Canada, Mexico, Peru and Italy (Serravezza, Sarrabus).

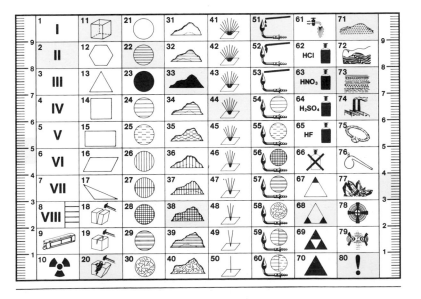

$Cu_3SbS_{3.25}$ Peru

Copper antimony sulphide. Can contain silver, mercury, arsenic, bismuth, nickel, tellurium, tin. It exists in many varieties: freibergite with silver, schwatzite with mercury, annivite with bismuth; frigidite with nickel, and goldfieldite with tellurium. It can be found in the same localities as tennantite. Beautiful crystals can be found at Kapnik (Hungary) and Botes (Rumania). It can be a source of precious elements, such as silver, mercury, tin, antimony and tellurium. Exposed to a flame it produces green flashes and white vapour.

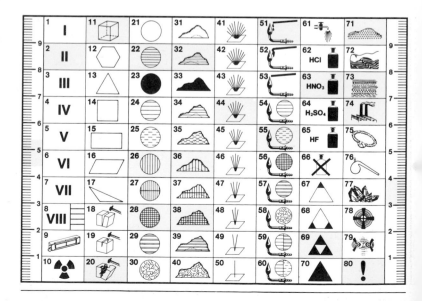

Cu_3AsS_4 Peru

Copper arsenic sulphide. Often in association with iron and zinc. Excellent source of copper. Beautiful prismatic crystals, sometimes tabular, granular masses. Can be found at Bor (Yugoslavia), Tsumeb (Namibia), Butte, Tintic and Bingham (USA) and Santa Fe mine (New Mexico). There are also numerous deposits in South America: Coquimbo, Atacamba, San Pedro Nolasco (Chile), Morococha, Cerro de Pasco, Quiruvilca (Peru) and Sierra de Famatina (Argentina). It has also been found at Luzon (Philippines). Antimony does sometimes substitute arsenic. Applied to a flame it produces green flashes and white vapour.

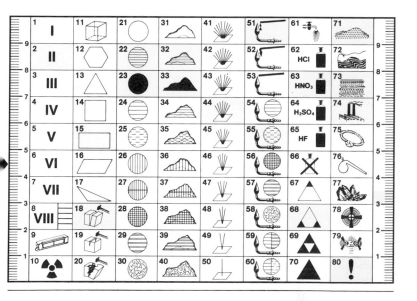

PbS   Kansas (USA)

Lead sulphide. The most diffuse and abundant lead mineral. The richest world deposits are found in the USA (Missouri), Canada (British Columbia), Mexico, Australia (Broken Hill, New South Wales), the USSR and India. Good deposits exist in Germany, Austria, Yugoslavia, England (Cornwall and Derbyshire) and Italy (Sardinia, northern Italy and the eastern Alps). Zinc minerals are usually found in association with galena deposits.

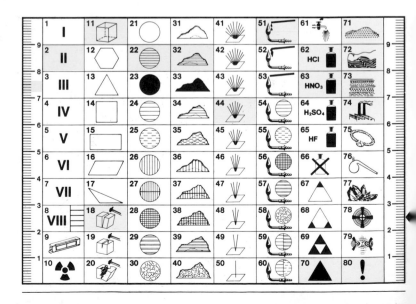

Sb_2S_3 Rumania

Antimony sulphide. Also called stibnite. Fusible with a match, emits white vapour. Beautiful crystals have been found in Rumania (Felsobanya); also present in E. Germany, Hungary, Czechoslovakia and central France. The world's largest deposit is in China (Hunan and Kwangtun); others are in Borneo, Bolivia, Peru, Mexico, USA (California, Idaho, Nevada) and Italy. Perfect crystals, as long as 45cm, have been found in the now exhausted mine of Shikoku (Japan). Apart from being the main source of antimony, it is used in the textile, rubber and glass industries.

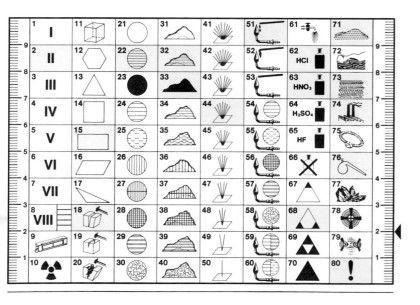

Bi_2S_3 Brosso (Italy)

Bismuth sulphide. Very similar to antimonite. Often associated with other minerals, it can contain lead, iron and copper. Large deposits can be found in Bolivia (San Baldamero, Potosi, Tasna, Chorole), Peru (Cerro de Pasco) and Mexico (Guanajuato); in the latter deposit selenium is often found to substitute sulphur. Splendid crystals have been found in California and Dakota, as well as in Italy (Brosso, Crodo, Boccheggiano, Elba and Sardinia).

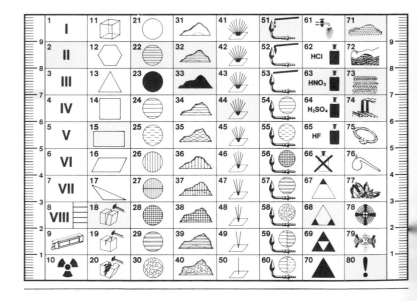

FeAsS   Yugoslavia

Iron arsenide sulphide. Can include cobalt, and originated 'glaucodote' (Co,Fe)AsS.
Often in twins or multiple twins. There are numerous deposits: Mitterberg (Austria),
Freiberg (E. Germany), Rozsnio (Hungary), Sao Jao Gonzao and Panasqueira
(Portugal), Boliden (Sweden), Sulitijelma (Norway), Belgium and Great Britain.
Beautiful samples have been found in the many deposits of the USA, Canada, Mexico,
Bolivia, and a few in Italy. It is the principal mineral source of arsenic, but can also yield,
as secondary products, gold, silver, tin and cobalt.

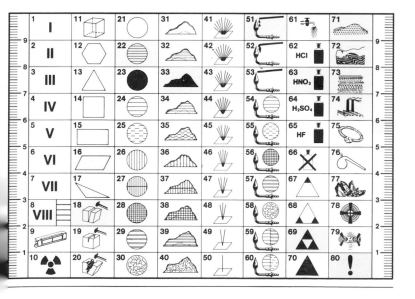

$2PbS.Cu_2S.Sb_2S_3$ Peru

Lead copper sulphantimonide. Part of antimony can be substituted by arsenic. Numerous German mines have produced it: Clausthal, Andreasberg, Wolfsberg, Wendorf (W. Germany); Braunsdorf and Altenberg (E. Germany). Splendid crystals have come from Liskeard (Cornwall, England) and good samples from Huttenberg (Austria), Pribram (Czechoslovakia) and Felsobanya (Rumania). In Spain it can be found in the Sierra Almagrera at Guadalajara, and at Almaden in association with cinnabar. It is always present in the silver and tin mines of Bolivia, Chile, Peru, Mexico, and in various deposits in the USA (Nevada, California, Arkansas, Utah and Montana), as well as Canada (Ontario) and Italy (Sardinia, Piedmont, Tuscany).

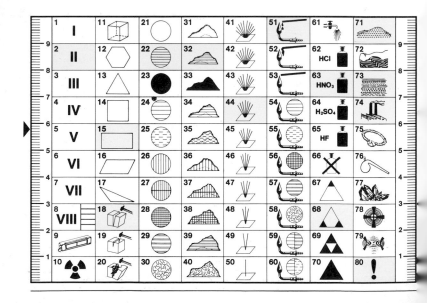

$CuPb_{13}(SbS_3)_7S_3$ Serravezza (Italy)

Lead copper sulphantimonide. Formed by long acicular crystals, vertically striate. A rare mineral, it was first discovered at Bottino, near Serravezza, in the Tuscan Appennines. Usually associated with blende, galena, chalcopyrite, jamesonite and boulangerite. Recently found in many lead-zinc deposits of hydrothermal origin: at Schwarzenberg (E. Germany), Hallefors (Sweden), Goldkronach (Bavaria, W. Germany), Marbe Lake (Ontario, Canada), Broken Hill (NSW, Australia).

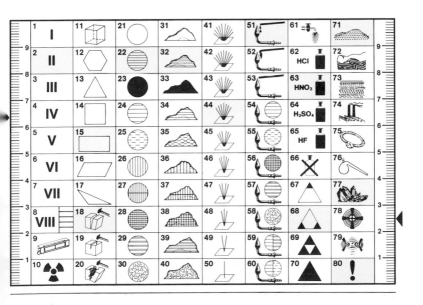

MoS$_2$ Traversella (Italy)

Molybdenum sulphide. Tubular crystals, more often in foliated masses. Valuable for its content of molybdenum, which is used in many alloys. Large deposits are found in Australia, USA, Canada and the USSR. It is an accessory of granitic and dioritic rocks and can also be found in metamorphic rocks. Certain quantities can be found in the quartz veins contained in the granites of Norway, Bolivia, Brazil and other areas, like Sardinia, Calabria, Varese (in porphyry deposits).

FeFe$_2$O$_4$ Traversella (Italy)

Ferrous iron oxide. Iron can be partly substituted by other elements, thus originating several varieties: titanomagnetite, chromomagnetite, nickel magnetite, manganomagnetite and vanadomagnetite. It is present worldwide and is the best iron ore. One of the largest deposits is the one at Kiruna (Sweden); some of the richest are those in the Urals (Magnetigora, Blagodat). It is also found in the Siebengebirge and Kaiserstuhl (W. Germany), Kraubat (Austria) and Oravicza (Hungary); in many US areas (Iron Springs, Iron Mountains, Adirondacks) and in some important deposits in Italy (Traversella, Elba, Sardinia).

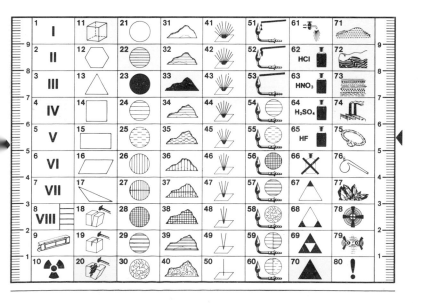

Fe_2O_3 St Gotthard (Switzerland)

Iron oxide. Squat rhombohedral crystals, thin plates or plates in the form of rosettes ('iron roses'), or red earthy masses (red ochre). It is the most abundant iron ore, often used as a substitute for magnetite, the apparent shape of which it often assumes thus originating the variety 'martite' (pseudomorphosis). Important deposits have been found at Krivoi Rog (USSR), Bilbao (Spain), Westphalia (W. Germany), Kragero (Norway) and Cumbria (England). Huge formations have been found at Lake Superior (USA). Also quite important are the mines in Canada (Quebec), Venezuela, Brazil and Angola. In Italy, Elba and Piedmont have produced extremely beautiful crystals.

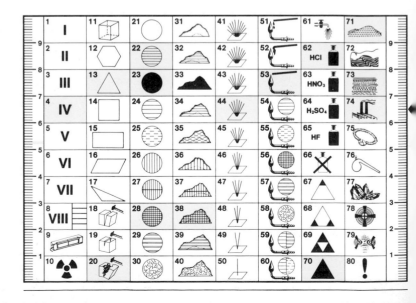

FeTiO$_3$

Norway

Iron titanium oxide. Highly resistant to atmospheric agents, ilmenite can be found in great quantities in sedimentary deposits such as those of Travancore (India). Large crystals (12–13cm in diameter) have been found in Kragero (Norway) as well as at Arendal, Ekersund, Snarum (also Norway). It is also found in Routivare (Sweden), in vast deposits in the Urals, Siberia, Australia and India (Kishangarh). Also quite important are the mines in Canada, Florida (USA) and Brazil, while splendid crystals have been found in northern Italy (Val Devero, Val Malenco, Val di Susa).

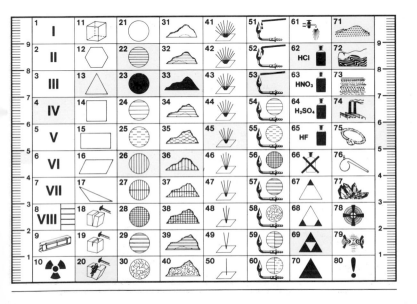

β-TiO₂   Norway

Titanium dioxide. This compound can be found in three polymorphic modifications: anatase, rutile and brookite. A rare mineral present in northern Italy, Norway (in considerable deposits), England (Cornwall), France, southern Urals, and Switzerland (Gotthard, Binnthal). Magnificent blue crystals have been found in Colorado and Massachusetts (USA). It is less rare in secondary sedimentary deposits.

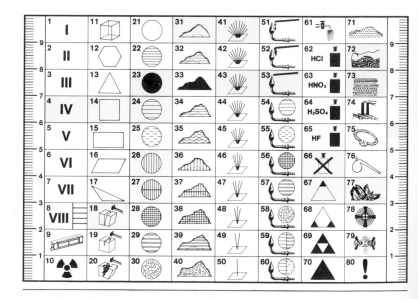

MnO.OH   Harz Mtns (W. Germany)

Basic manganese oxide. Typical mineral of low temperature hydrothermal veins. Chemical alterations can lead to pyrolusite, psilomelane, braunite and hausmannite. The best crystals have been found in Ilfeld (Harz), where they were up to 8cm long. Manganite is also found at Ilmenau (E. Germany), the High Pyrenees, Egremont and Exeter (England). Vast deposits of secondary, sedimentary origin have been found in the Ukraine; the USA and western Canada also possess several deposits. It could be an excellent mineral ore for manganese, but it seldom appears in sufficient quantities.

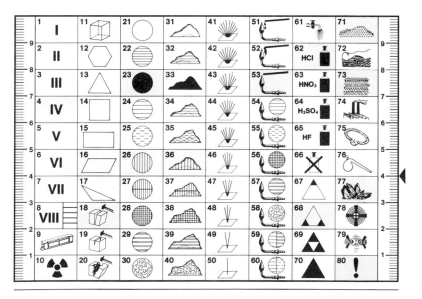

Au   California

Gold. Rare in crystals. Frequent as scales and granules. Its yellow colour varies slightly according to the impurities it contains. In association with silver it produces electrum; with palladium, porpezite; with rhodium, rhodite. Its specific gravity is extremely high, 19.3, and it melts at 1061°C. It is native of quartz veins, and, after their disintegration, can be found in fluvial deposits. The main gold deposits are to be found at Witwatersrand (South Africa), Yukon (Alaska), Porcupine (Canada), the Urals and Siberia. Small quantities are extracted from the quartz veins of the Alps, and the arsenopyrites and gold sands of northern Italy.

19,3

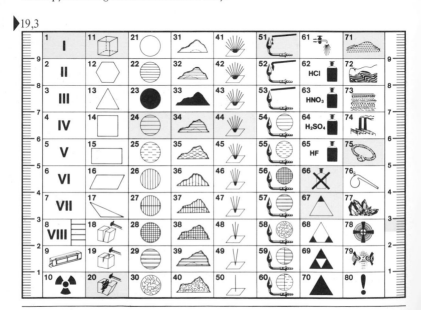

S-α   Agrigento (Italy)

Sulphur. Melts at 119°C. Has its origin in the sublimation of sulphates present in water. It is often associated with oil deposits and sublimates in vast quantities during volcanic eruptions. The most important deposits are found in the USA (Texas and Louisiana), Japan, and Italy (Sicily and Romagna). Large quantities of sulphur are today extracted from oil. Crystals are quite frequent in bi-pyramidal forms. It is susceptible to the slightest thermal variations, which produce internal tensions and fractures.

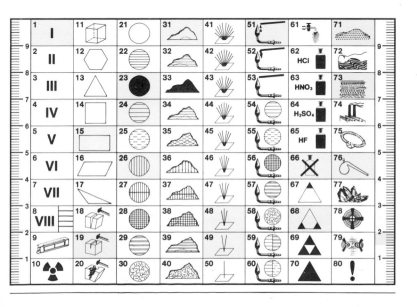

CuFeS$_2$ Oklahoma (USA)

Copper iron sulphide. The most common and important mineral for the extraction of copper. Large deposits are found in Spain (Rio Tinto, Huelva), Czechoslovakia (Schemnitz), Austria (Kitzbuehl), E. Germany (Freiberg), W. Germany (Harz Mtns), France (Alsace, Isère), Sweden (Falun), Chile (Braden), Peru (Maracacha), and Mexico (Los Pilares). Other important deposits are those of the USA (Butte, Montana), Canada (Sudbury), Zaire, Zambia, the Urals, Cyprus, Tasmania and Australia. Also present in northern Italy, Liguria, Tuscany and Sardinia.

CdS   Monteponi (Sardinia)

Cadmium sulphide. Extremely rare mineral which forms within zinc deposits as a result of alterations in the blende. In this sample, a thin layer of microcrystals covers dolomitic rhombohedrons. It has been found in the labradorite porphyry caves of Bishopton (Scotland) in association with calcite, prehnite, natrolite; at Bleiberg (Austria), Pribram (Czechoslovakia), Laurium (Greece) and various localities in the USA. Its interest is exclusively scientific (research, collections, etc).

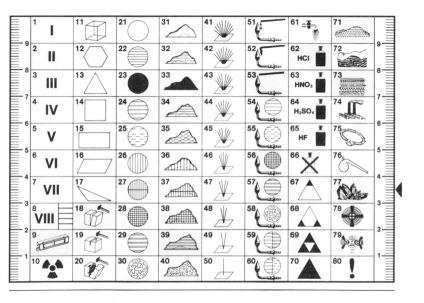

$Fe_{11}S_{12}$ Yugoslavia

Iron sulphide. Also called magnetopyrite, it has magnetic properties. Its exploitation is not economical, but its frequent association with pentlandite $(Fe,Ni)_9S_8$ makes it the most useful mineral for the extraction of nickel. Excellent crystals have been found in Yugoslavia (Trepca), Rumania (Kisbanya), Austria (Leoben), W. Germany (Saxony) and W. Germany (Bavaria). Huge deposits occur in Canada (Sudbury, Manitoba), the USA, Mexico, Bolivia and Brazil. Minor ones are present in Italy (Piedmont: Val d'Ossola, Val Sesia, Traversella; Tuscany: Serravezza, Elba).

NiS   Transvaal (S. Africa)

Nickel sulphide. Acicular radial crystals, elastic but fragile. Brassy yellow in colour. It originates in low temperature lithoclases. Can also derive from the alteration of other nickel minerals. Only soluble in turpentine. It has been found in small quantities in Johangeorgenstadt (E. Germany), Andreasberg and Saarbrücken (W. Germany), Emilia and Sardinia (Italy), in some mines in Canada (Sudbury), the USA (Iowa, New York), South Africa, and, above all, in Wales. Although rich in nickel, its scarcity makes it uneconomical to exploit.

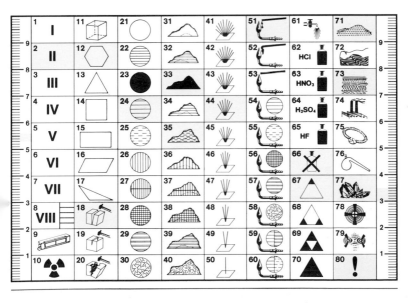

$(Co, Ni)_3S_4$ Missouri (USA)

Cobalt nickel sulphide. A rare mineral belonging to the linnaeite group. It was first found at Siegen (W. Germany). It occurs in the deposits of La Matte (Missouri), and, sporadically, in various mines of nickel sulphides (Canada). Its interest is purely scientific. Its colour is a yellowish steely grey, its lustre metallic and vivid. Crystals are mainly minute octahedrons. It melts easily. Besides siegenite, the linnaeite series includes some ten rare species of sulphides.

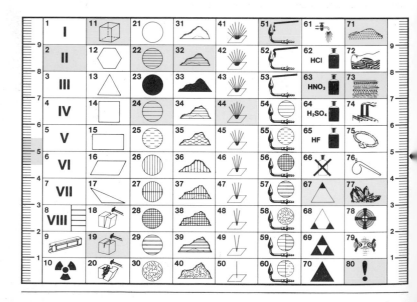

FeS$_2$ Grosseto (Italy)

Iron disulphide. The most widely spread and abundant sulphide in the lithosphere. Of great interest to industry, museums and collectors who are attracted by its splendid crystals, usually large and variously shaped. Deposits are so numerous it is impossible to list them all. The best known ones are: Rio Tinto (Spain), Falun (Sweden), Sulitjelma (Norway), Gelman, Chester, Schoharie and Sparta (USA). Well known throughout the world are the splendid pentadodecahedrical pyrites of Elba (Rio Marina), the cubic triglyphs from Gavorrano (Grosseto, Tuscany), the octahedrical crystals from Brosso and Traversella (Piedmont, Italy). Peru also produces pyrites of quite exceptional lustre. Pyrites are used to extract iron, gold, copper, nickel, cobalt; they are also indispensable in the production of sulphuric acid.

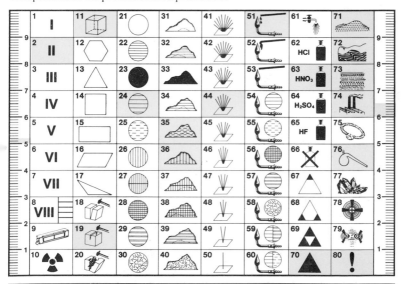

FeS_2 Spain

Iron disulphide. Can contain small quantities of arsenic, antimony, and thalium. It occurs in curious forms—the crystals are twinned in unusual, complex units: 'spears', 'cockscomb', discs, spheres and bunches. Good crystals have been found at Littnitz and Karlovy Vary (Czechoslovakia), Rumania, Rammelsberg (Germany), Cap Blanc-Nez (France), Dover and Tavistock (England), in the mining district of the Three States (USA) and Sardinia (Silius). If found in quantity, it is used for the production of sulphuric acid; it can also be used to imitate antique jewellery.

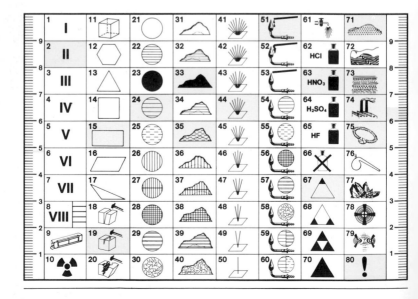

SiO$_2$ Argentina

Silicon dioxide. A micro-crystalline variety of quartz. It usually appears with veins and presents a fibrous and micro-granular structure. It is subdivided into many varieties: 'agate' when the veins are concentric and not too contrasting; 'onyx' if the contrast is higher; 'carnelian' if the veins are red due to the presence of ematite; 'chrysoprase' if it is tinged apple-green by traces of nickel; 'enhydros' if it contains liquid remains. Large quantities of chalcedony, used for ornamental purposes, come from Brazil, Uruguay, India, the Urals, and Australia. Its wide distribution makes it economical to extract.

Carnotite (c. ×1.5)

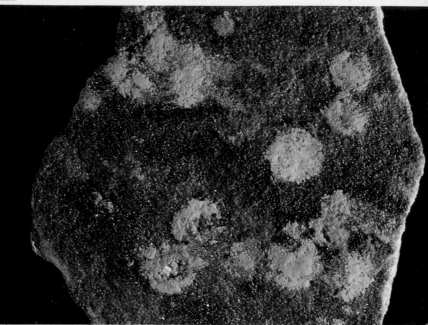

$K_2[(UO_2)_2(VO_4)_2],3H_2O$ South Dakota (USA)

Hydrated potassium uranium vanadate. One of the main minerals of uranium and vanadium. Widespread in the south-west of the USA, where it is used in the production of uranium. Contained as dust or lenses in Jurassic sandstones. It is thought to be of secondary origin, as the result of the action of meteoric waters on pre-existing uranium minerals. The most important deposits are at Paradox Valley (Colorado), Utah, Arizona, New Mexico and Nevada; in the desert of Fergana (USSR), in some sandstone veins in Katanga and Zimbabwe (Africa), and at Radium Hill (southern Australia). It has no lustre.

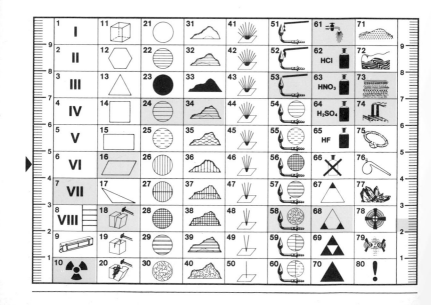

$Ca_2(Al,Fe''')Al_2(O|OH|SiO_4|Si_2O_7)$ Monte Rosa, Aosta Valley (Italy)

Basic calcium aluminium ferric iron silicate. The variety 'piedmontite' contains manganese; 'allanite' includes various rare earth elements. The best crystals are found in the lithoclases of crystalline schists. It is a typical collector's mineral, seldom used for jewellery. Splendid crystals come from Krappenwand (Austria), Arendal (Norway), Bourg-d'Oisans (France), Zermatt (Switzerland), the Urals and Sulzer (Alaska). The epidotes found in California (Riverside), Massachusetts (Woburn) and Connecticut (Haddam) are well known; some good ones come from Elba and the Piedmontese Alps (Italy).

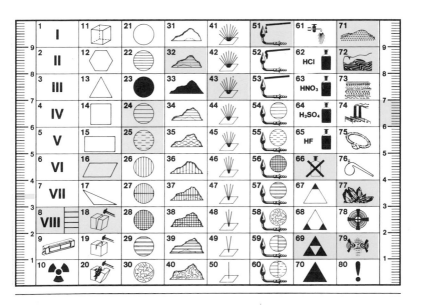

Cu_5FeS_4 France

Cuprous iron sulphide. Also called 'purple copper' because of its iridiscent colour. Well-formed crystals are rare. Usually occurs as massive, compact or granular. In Europe it can be found in E. Germany (Freiberg and Mansfeld), in the Ardennes, in Norway and in Sweden. Beautiful crystals have come from Pragratten (Austria) and Redruth (Cornwall, England). It occurs in many copper deposits at Butte and Bristol (USA), Mexico, Peru, Chile, Tsumeb (Namibia), Zambia and Australia. It can be found in small quantities in Sardinia.

$SiO_2.nH_2O$ Australia

Hydrated silicon dioxide. It occurs in an amorphous or microcrystalline state with a structure similar to cristobalite. The variety hyalite is pure opal, colourless and clear. Beautiful colours can be created by the incidence of light on its clusters of particles. 'Precious opal' is the variety with exquisite iridescent colours (New South Wales); 'fire opal' has richly red 'internal' lights (Queretaro, Mexico); 'wood opal' produces the trunks of the 'petrified forests', in which the silica has replaced the organic parts while reproducing their shape and structure. Classic deposits are found in Czechoslovakia, the USA, Mexico, Honduras and Australia.

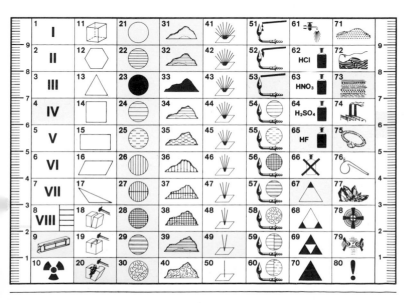

$Cu_3(OH|CO_3)_2$ Tsumeb (Namibia)

Basic copper carbonate. It is a mineral of secondary origin, which forms in the oxidation zone of copper deposits following the action of waters rich in CO_2. It occurs in association with various copper minerals, such as chrysocolla, chalcocite, and above all malachite, into which it slowly reverts. It is widespread but not in quantity. Excellent crystals have been found at Chessy (France), Tsumeb (Namibia), Laurium (Greece), Morocco, and various mines of Arizona. It occurs in the copper deposits of Chile, Mexico, the Urals, Siberia and Australia (Broken Hill), where interesting crystal formations have been found. Good samples have come from Sardinia and Tuscany (Italy).

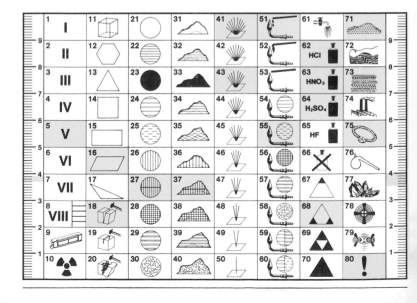

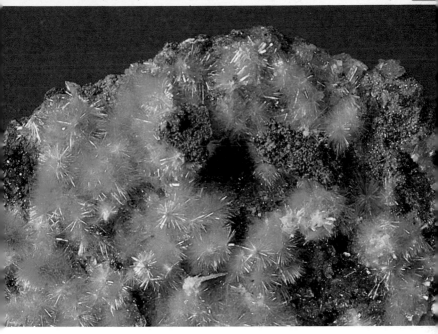

$(Zn,Cu)_5[(OH)_3CO_3]_2$ Arizona (USA)

Basic zinc copper carbonate. A mineral with purely scientific interest, once called 'brass flowers', extremely rare in isolated crystals. Almost always found in efflorescences of acicular crystals or thin crusts. It is of secondary origin, since it formed in the oxidation areas of zinc and copper. The best samples came from the mines of Bisbee (Arizona), Mexico, Cumbria and Derbyshire (England). In Italy some have been found in the zinc deposits of Sardinia and the Alps.

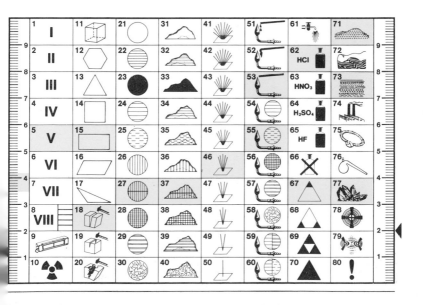

$Sr(SO_4)$ Malagasy Republic

Strontium sulphate. It occurs in very beautiful crystals, colourless, white and light blue. Often a compound of barium sulphate and calcium sulphate. Beautiful samples have been found worldwide: Malagasy Republic, Bristol (England), Dornburg (E. Germany), Herrengrund (Czechoslovakia), Leogang (Austria), Bex (Switzerland), Tunisia and Turkestan. Large crystals of over 0.5m in length and 2–3kg in weight, white or opaque, were found at San Bernardino and Put-in Bay (USA). The museums of the world display Sicilian celestines associated with sulphur, calcite, aragonite and rock-salt, but also beautiful are those found in the basalt caves near Vicenza (Italy).

PbCu[(OH)₂SO₄]   New Mexico (USA)

Basic lead copper sulphate. The name comes from Linares, Spain, where it was first found. It occurs in elongated, tabular crystals or acicular groups. It is a secondary mineral of the oxidation zone of lead and copper deposits. Crystals up to 10cm long have been found in Arizona; beautiful ones come from the deposits in Tsumeb (Namibia), from the Sierra de Capillatas (Argentina), Leadhills (Lanarkshire), Broken Hill (Australia) and above all from the deposits in Utah (Tintic), Montana (Butte) and California (Cerro Gordo Mine). Samples have also been found in the mines of Sardinia.

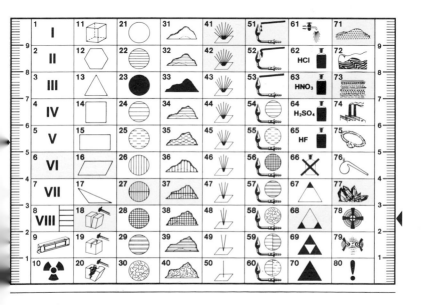

$(Mg,Fe'')Al_2(OH|PO_4)_2$ Yukon (Canada)

Basic phosphate of iron, aluminium and magnesium. It forms an isomorph amalgam with scorzalite: if magnesium prevails on iron, lazulite occurs; if iron prevails on magnesium, the isomorphic sequence produces scorzalite. Good crystals have been found at Werfen and Graz (Austria); larger ones at Horrsjoberg (Sweden). It occurs in Brazil (Corrego Frio, Minas Gerais), Bolivia, Malagasy Republic, Canada and the USA (Georgia, Carolina). Splendid samples have recently been found in the Yukon. It is often in association with rutile, corundum, andalusite and cyanite. It is used for ornamental purposes but is not very important in this respect.

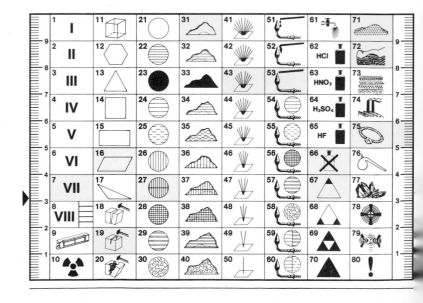

$Ca_5[(F, Cl, OH) (PO_4)_3]$ Lake Baikal (Siberia)

Calcium phosphate in combination with various anions. If it contains fluoride it is called fluor-apatite (the most common variety); if chlorine, chlor-apatite; if hydroxides, hydroxyl-apatite. It occurs in variously-coloured beautiful crystals, often quite large. The 'phosphorites' are deposits of calcium phosphates with organic origin, used as fertilizers. Green apatites have been found at Wilberforce (Ontario), extremely clear yellow ones at Durango (Mexico), purple ones at Ehrenfriedersdorf (E. Germany) and blue ones in Siberia, Knappenwand (Austria) and Auburn (Maine, USA). Clear yellow crystals have also been found in the ejecta of Vesuvius and in the Piedmontese Alps in Italy, while other alpine regions have yielded greenish-grey and opaque white apatites.

$CuAl_6[(OH)_2PO_4]_4 \cdot 4H_2O$ Mexico

Hydrated basic phosphate of aluminium and copper. An ornamental stone, extensively used since antiquity, and brought to the West by the Turks (hence its name). One of the classic sources of turquoise, where splendid blue samples have been found, is Nishapur (Iran). Equally famous is the Sinai area, the Magara valley in Saudi Arabia, and the Samarkand district in Turkestan. Excellent nodules have been found in Nevada, New Mexico and Arizona (USA). It is a mineral of secondary origin, very rare in crystal forms, occurring in dry climates and aluminium-rich rocks. The less precious varieties are often artificially coloured, and some other minerals are disguised to look like turquoise.

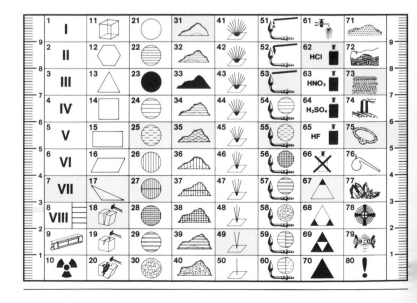

$Al^{(6)}Al^{(6)}(O|SiO_4)$ Brazil

Aluminium silicate. The name is derived from its blue colour (gr. *kyanos*). Abbot Hauy also called it disthene as it is extremely hard in the direction orthogonal to the length of the crystals and half-hard in the parallel direction. It occurs in metamorphic rocks and it is invaluable for the study of the metamorphic process. Excellent blue kyanites can be found at Pizzo Forno (Switzerland); highly esteemed are those of Kashmir and Burma, now used for ornamental objects; the green ones of Machakos (Kenya) are quite unique; there crystals reach 30cm in length. Kyanites are used in the production of porcelain.

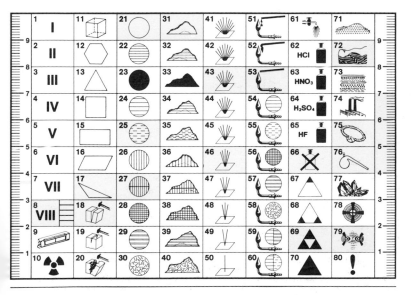

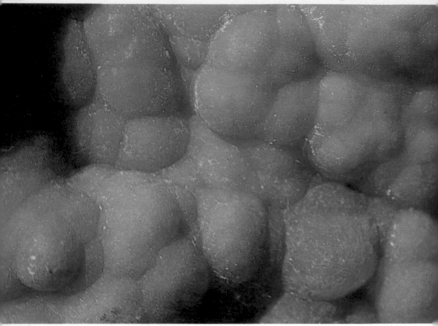

$Zn_4[(OH)_2Si_2O_7].H_2O$ Sardinia

Basic hydrated zinc silicate. Called calamine by miners; its mineral ores are amalgams of hemimorphite, smithsonite and hydrozincite. When it contains copper it occurs in a splendid blue colour; excellent samples thus coloured came from the mines of the Iglesiente (Sardinia), now worked out. It is a secondary product of the oxidation zones of lead and zinc deposits. Almost always in association with smithsonite, galena, cerussite, anglesite and sphalerite. Deposits can be found at Vieille Montagne (Belgium), Aachen (W. Germany), Silesia (Poland), Pennsylvania, Montana, Colorado and New Mexico (USA), Siberia, Yugoslavia, parts of the Alps, Carinthia (Austria) and Matlock (England).

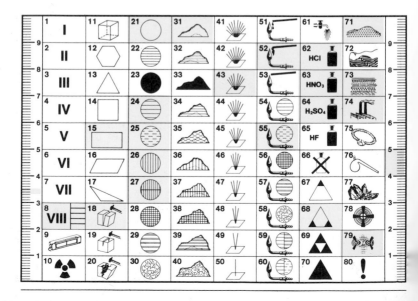

$Cu_4H_4[(OH)_8|Si_4O_{10}]$ Capo Calamita (Elba)

Hydrous copper silicate. Generally occurring as finely fibrous encrustations, or earthly masses, microcrystalline or stalactitic; also globular. Found in the oxidation area of copper deposits in association with malachite and azurite, it is due to the reaction of siliciferous waters with copper ores. Its presence indicates copper deposits. Large quantities have been found in Chile and in the USA (Pennsylvania, Michigan, Arizona, New Mexico, California, Idaho, Nevada, Utah, Montana and Colorado); as well as in the USSR, Katanga (Africa), Zaire, Sardinia and northern Italy.

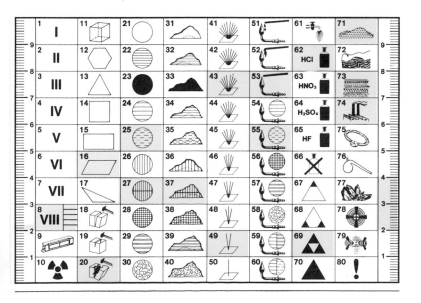

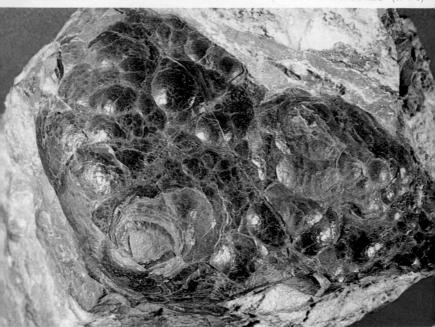

As   Brosso (Italy)

Native arsenic. Rarely as crystals (small and ill-shaped), almost always in concretions. Exposed to heat, it does not melt, but burns emanating white vapour (As_2O_3) with a characteristic garlic smell. It is not used as it is extremely rare; arsenic used in industry is a sub-product of the processing of arsenopyrites. Native arsenic has been found in Germany (Harz and Erzgebirge mtns), Siberia (Gikos), France, Rumania, Czechoslovakia, Japan and Italy (Piedmont).

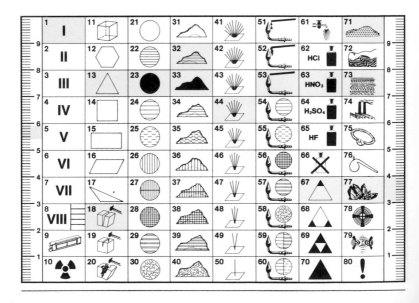

SnO_2 Bolivia

Tin dioxide. It occurs in short, twinned crystals, or as fibrous or concretionary masses, or as granular aggregates. An extremely important ore for the extraction of tin. Good quantities come from Cornwall (England), Saxony (E. Germany) and Bohemia (Czechoslovakia). More modest deposits are found in Spain and Portugal. Tantaliferous cassiterite is found in France (Montebras). The most important deposits have been found in Malacca, Sumatra and Bolivia; good ones exist in the USSR, China, the USA and Canada. The Italian deposit of Monte Valerio (Tuscany) was mined by the Etruscans.

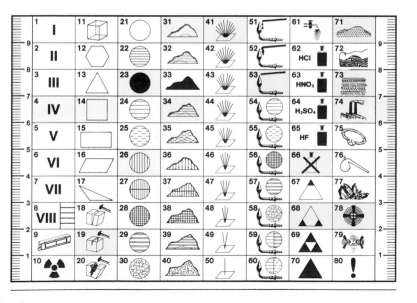

(Mn,Fe)(WO₄)   Portugal

Iron manganese tungstate. It is the isomorphic alloy of two mineral ores: ferberite $FeWO_4$, and huebnerite, $MnWO_4$. It forms large tabular crystals vertically striated. It is very important as the source of wolfram or tungsten, occurring in hydrothermal veins with cassiterite, molybdenite, pyrite, or in granite with micae, tourmalines and fluorite. It has been found in the Erzgebirge mtns (E. Germany) and Cornwall (England); in alluvial deposits in Spain and Portugal, and in considerable quantities in southern China, Colorado (USA), and Bolivia.

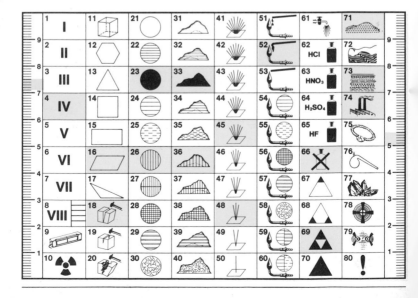

MnWO$_4$ Peru

Manganese tungstate. Beautiful elongated, prismatic crystals, often flattened, in parallel groups. They vary in colour from yellow to bright red, brown and black. With wolframite and ferberite, it is one of the so-called 'series of wolframite': all three minerals are the main source of tungsten. Huebnerite occurs in granite and hydrothermal veins of high and medium temperature. Excellent crystals are found especially in Peru and various localities in the USA: Colorado (particularly Uncompahgre, near Gladstone, and Silverstone), Montana (Butte) and New Mexico (White Oaks).

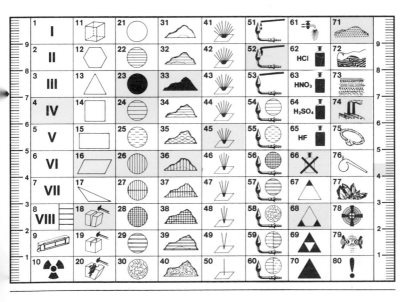

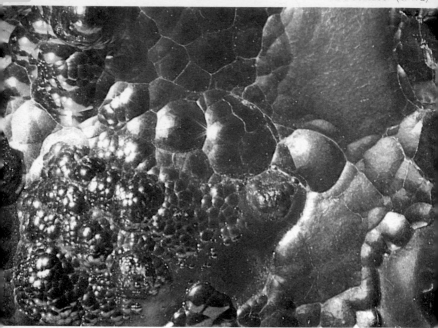

FeOOH   Elba

Basic ferrous oxide. Very rare as crystals, it usually occurs as acicular aggregates, radiating scales, botryoidal, granular or earthy masses, and in stalactite-stalagmite formations. It constitutes a great part of the common limonite, originated by the oxidation of iron ores and found as deposit in alluvial basins ('swamp iron'). Goethite crystals have been found at Pribram (Czechoslovakia), in Cornwall (England), the USSR and Colorado (Pikes Peak). Large deposits exist in Alsace-Lorraine, Westphalia (W. Germany), Cuba and Labrador (Canada). It is used for the extraction of iron and colouring ochres.

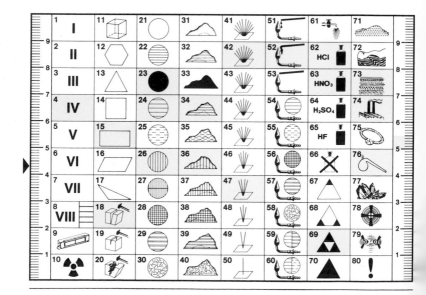

$(Mg,Fe'')_2Fe'''(O_2|BO_3)$ Brosso (Italy)

Magnesium iron borate. A prevalence of magnesium produces ludwigite, while a prevalence of iron gives origin to vonsenite. These two minerals are the extreme members of the ludwigite series. Ludwigite crystals are acicular, in radial fibrous masses. It has been found in Banato (Rumania), in the magnetite deposits at Norberg (Sweden), and in Idaho, Nevada and Montana (USA). Typical forms of vonsenite come from California (Riverside). Excellent samples of ludwigite associated with magnetite and as charite have come from the mines of Brosso; vonsenite has been found in the region south of Naples.

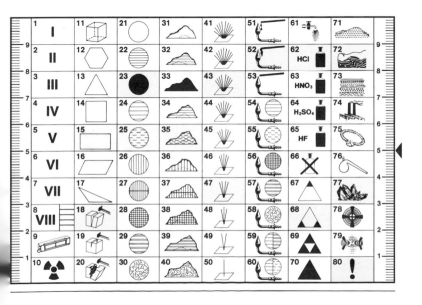

$CaFe_2''Fe'''(OH|O|Si_2O_7)$ Elba

Basic calcium iron silicate. It occurs in prismatic crystals, more or less elongated and vertically striated. It was first found on Elba, hence its name (*Ilva* was the island's Latin name). It has purely scientific or collectors' interest. It has been found at Trepca (Yugoslavia), Seriphos (Greece), Siorsuit and Kangerdluarsuk (Greenland), Laxey (Idaho, USA) and in the Urals. Splendid crystals come from Rio Marina and Capo Calamita (Elba), and other parts of Tuscany and Sardinia. It is often in association with magnetite, hedenbergite and quartz.

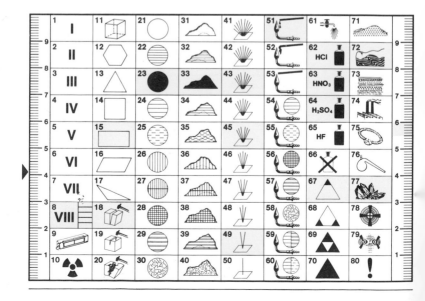

$NaFe_3'''Al_6[F|O_3(BO_3)_3Si_6O_{18}]$ California (USA)

Cyclosilicate compound containing boron. A rare mineral, first found near Mexquitic (Mexico) in 1966. It was given its name in honour of Professor M. J. Buerger, of the Massachusetts Institute of Technology, who first defined its structure. It occurs in dark brown, prismatic crystals, sometimes almost black, and opaque. It belongs to the tourmaline group, from which it differs in its composition and its distinct prismatic cleavage. It also has a bronze lustre. The sample above comes from Boleo (California).

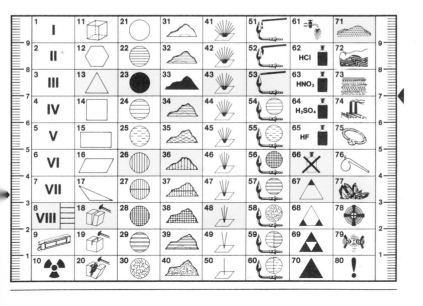

$(Ca,Na,K)_{2-3}(Mg,Fe'',Fe''',Al)_5[(OH,F)_2(Si,Al)_2Si_6O_{22}]$ Monte Rosa (Italy)

Complex aluminosilicate compound belonging to the group of monoclinic amphiboles. It occurs as prismatic crystals, radial acicular crystals, or massive. It comprises many varieties: common hornblende, rich in bivalent iron, with a dark green colour; basalt hornblende, darker and rich in trivalent iron; a pale green hornblende, with little iron, typical of the calcareous contact rocks of Edenville (New York), Pargas (Finland) and Kotaki (Japan); the very rare tschermakite found at Pernio (Finland); and the hastingsite of the gabbroes of Ontario (Canada), Norway, the Urals and parts of the USA.

NaCa$_2$Mg$_4$(Al,Fe''')[(OH,F)$_2$AlSi$_6$O$_{22}$]   Pargas (Finland)

Complex aluminosilicate, one of the amphiboles. It occurs either as thick tabular crystals or as compact, foliated masses. It is dark brown or blackish. It originates within contact-metamorphic limestones or femic intrusion rocks. The pargasite found at Pargas is well known; other deposits have been discovered at Mansjo and Langban (Sweden), in the gabbro of Burlington (Pennsylvania), in the peridotites of Tinaquillo (Venezuela) and in the seamy rocks near Sondrio (Italy). The 'carintine' variety is common in the eclogites of Saualpe (Austria), Nowa Wies (Poland) and Münchberg (W. Germany).

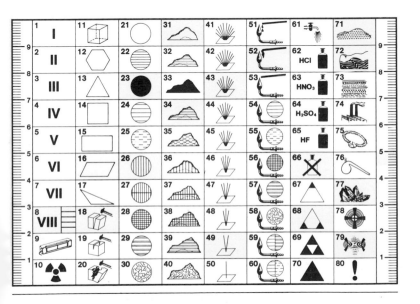

$KNa_2Li(Fe,Mn)_2Ti_2(O|SI_4O_{11})_2$ California (USA)

Silicate of sodium, iron and titanium. An extremely rare mineral with slender, shiny, prismatic crystals, deep black or dark brown in colour. Splendid samples come from San Bernardino (California) in association with benitoite and natrolite. Good ones have been found in the nephelinic syenites of the Kola peninsula (USSR); it has also been found in the foyatic pegmatites of Greenland and Ireland. It has purely scientific interest and is much appreciated by collectors.

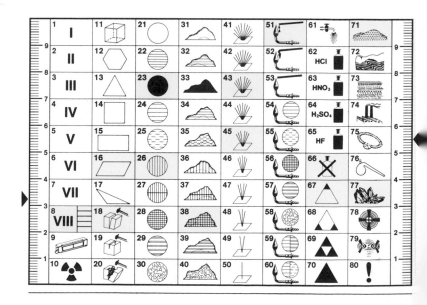

$K(Mg,Fe, Mn)_3[(OH.F)_2AlSi_3O_{10}]$ Vinadio (Italy)

Complex phyllosilicate of potash, magnesium, iron and manganese. Biotite can include other elements or vary the quantity of its components, thus originating different varieties: titanobiotite (with titanium), manganophyllite (rich in manganese), lepidomelane (containing iron). It turns to chlorite when exposed to atmospheric agents. Abundant worldwide occurrence: igneous and metamorphic rocks, or as a secondary mineral in sandstone deposits. Huge scale deposits have been found near Lake Baikal and Mount Ilmen (USSR), in Greenland, Scandinavia and Brazil. Beautiful crystals have been found in the ejecta of Vesuvius and in northern Italy.

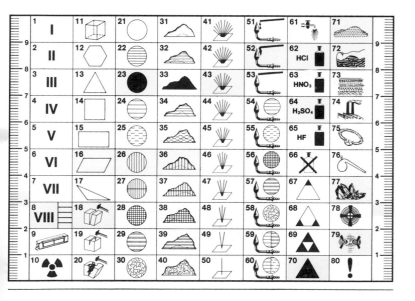

$Fe''_4Fe_2'''[(OH)_8Fe_2'''Si_2O_{10}]$ Kisbanya (Rumania)

Phyllosilicate of bivalent and trivalent iron with hydroxyls. Crystals variously hexagonal, trigonal, monoclinic, usually vertically striated; also fibrous, and reniform groups or thinly foliated. Its colours range from coal black to dark brown. Its dust is a dark olivaceous grey. Cronstedtite has exclusively scientific interest; it is an accessory of metallic ores of hydrothermal origin. It has been found, in association with limonite and chalcite, in the silver ore veins of Pribram and Kuttenberg (Czechoslovakia); it also occurs in England (Cornwall) and Brazil (Congonhas do Campo).

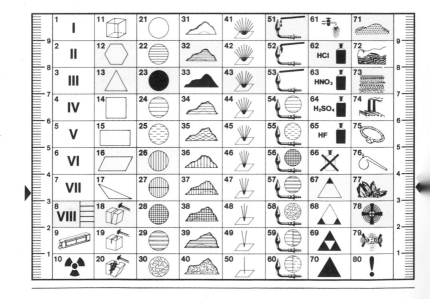

$Cu_2(OH)_3Cl$ Alacama (Chile)

Basic copper chloride. It results from superficial alterations of copper deposits, in arid climates, due to oxidation. Where it is found in quantity, eg Chile, it is used for the extraction of copper. It can be distinguished from malachite by the fact that, in hydrochloric acid, it does not produce effervescence, and from bronchantite by the fact that the flame it is exposed to assumes a green colour without injection of hydrochloric acid. Large crystals have been found in Australia (Ravensthorpe, Wallaroo), and good samples come from the copper deposits of Chile, Peru, Bolivia, Mexico, Bisbee and Jerom (Arizona), Tintic (Utah), the Urals, and the lava encrustations of Vesuvius and Etna.

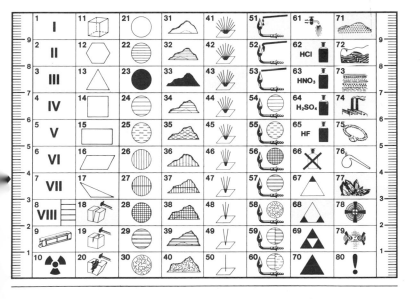

α-SiO$_2$ Elba

Silicon dioxide. Quartz is one of the most common minerals. It occurs in many different colours due to the inclusion of other minerals, natural radioactivity or heat. It is 'smokey' when brown to black, probably as the result of radiation; 'blue' if it includes particles of rutile or tourmaline; 'citrine' when it contains iron hydroxides; 'amethyst' if violet; 'pink' or 'rose' due to traces of manganese or titanium; 'maidenhair' with inclusions of yellow-red needles of rutile; 'aventurine' (green or brownish-yellow) if it includes scales of mica or goethite; 'tiger's eye' when it has inclusions of crocidolite; 'cat's eye' if it has asbestos fibres. The richest areas are Brazil, Uruguay, Malagasy Republic, the USSR and the Alps.

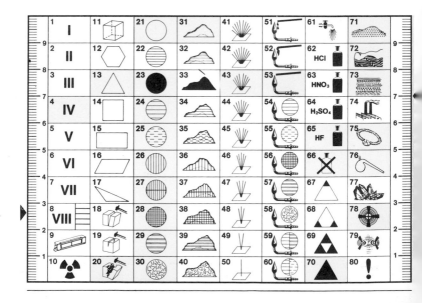

$Mg(OH)_2$ Vicenza (Italy)

Magnesium hydroxide. Usually found as foliated masses, rarely as crystals. Vivid pearly lustre on cleavage surfaces. It is the result of hydrothermal deposits in serpentine rocks with pyroaurite, hydromagnesite, brugnatellite, etc. The greenish-white variety occurring in long fibres is called nemalite. Extraordinary crystals, up to 15cm in diameter, have come from the mines of Lancaster (Pennsylvania); large deposits of nemalite exist in Asbestos (Quebec, Canada) and Baienov (USSR). Brucite associated with asbestos can be found in various parts of northern Italy.

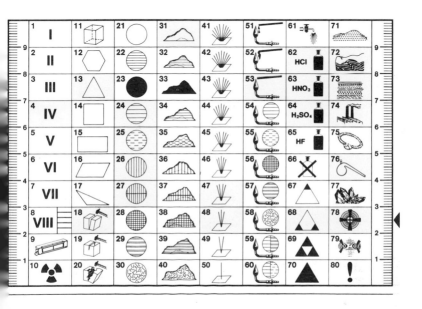

ZnCO$_3$ Tsumeb (Namibia)

Zinc carbonate. Isomorphic with the carbonates of the calcite series, so that part of its zinc content can be substituted with bivalent iron, calcium, manganese, magnesium, cobalt and copper. It is therefore an allochromatic mineral. It occurs in microcrystalline aggregates or concretions, rarely in well-formed crystals. It forms as a reaction of a zinc sulphate solution on limestones. Important deposits are found at Laurium (Greece), Santander (Spain), Vieille Montagne (Belgium), Bleiberg (Austria), W. Germany (Wiesloch), Poland (Beuthem), Rumania, Turkey, Mexico, the USA (Arkansas), Zambia, Namibia and Italy (Sardinia, and near Udine).

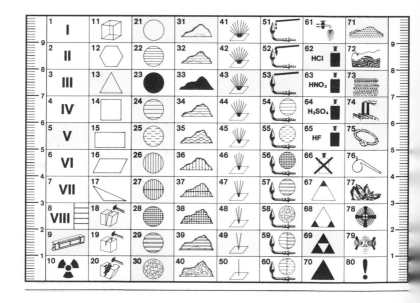

$Cu_2[(OH)_2CO_3)]$ Katanga (Zaire)

Basic copper carbonate. Rare in acicular crystals, it is usually found as fibrous-radial or zonal masses, and in stalactitic or botryoidal forms. Large blocks, weighing over 50 tons, were extracted during the last century from the Urals. Malachite deposits are worked in the Urals, Katanga (Zaire), Zambia, Chile, Australia (New South Wales), Tsumeb (Namibia) New Mexico, Utah and Arizona (USA). Splendid, well-known crystals have been found in parts of Europe: Chessy (France), Rezbanya (Rumania) and Betzdorf (W. Germany).

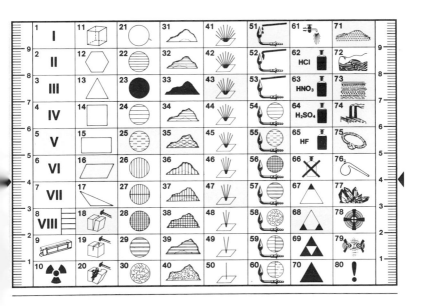

$Bi_2(O_2|CO_3)$ Mexico

Bismuth oxicarbonate. Widely spread, but not in great quantities, it is formed by the alteration of bismuth minerals. It occurs as earthy masses, granular or fibrous aggregates, rarely as good crystals. Normally grey or yellow, can become blue owing to the inclusion of copper compounds. It is found in Spain (Cordoba), E. Germany (Saxony), the Urals (Beresov), but the best known deposits are those in Bolivia (Tasna, Llallagua, Potosi, La Paz), Peru (Cerro Pasco) and Mexico (Guanajuato). It is also present in various US localities: Pala (California), Petaca (New Mexico), Nevada, Colorado, Arizona; and in Australia and Tasmania.

$Zn_2(OH|AsO_4)$ Mexico

Basic zinc arsenate. Isomorphic with olivenite (basic copper arsenate), with which it forms a continuous series with absolute interchangeability between copper and zinc. Its usually small crystals are grouped in fan shapes. Colours vary from brilliant green to yellow, pink or violet. Of interest only to the scientist and collector. It is a secondary mineral occurring in the oxidation area of zinc deposits. Deposits famous for their splendid samples are at Mapimi (Mexico), Chanarcillo (Chile), Tsumeb (Namibia), Laurium (Greece), Tintic (USA), Algeria, Turkey and Cap Garonne (France). In Italy it is associated with scorodite in the tin deposit on Monte Valerio (Tuscany).

$NaAl_3[(OH)_2PO_4]_2$ Minas Gerais (Brazil)

Basic sodium aluminium phosphate. A rare mineral discovered in 1945 by Pough and Henderson in the pegmatites of Conselheiro Pana (Minas Gerais), in association with muscovite, albite, apatite, lazulite and tourmaline. Its crystals can reach 6–7cm in length and present several faces. When clear, it can be used in gemmology. It was subsequently found in other Brazilian pegmatites (Pietras Lavadas) and, with small crystals, in the USA (Palermo Mine, Smith Mine in New Hampshire). It is of interest only to scientists and collectors.

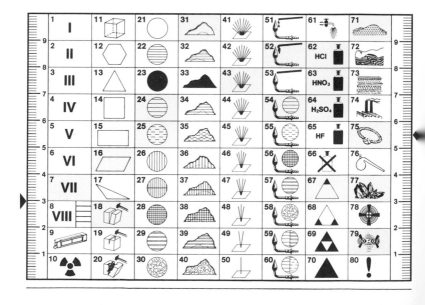

$Pb_5[Cl(PO_4)_3]$ Australia

Lead chloride phosphate. It is quite common as a secondary product of the oxidation zones of lead deposits, where it is found together with cerussite, emimorphite, smithsonite, anglesite, malachite, leadhillite and wulfenite. It forms isomorph amalgams with mimetite (lead chloride and arsenate). Pyromorphite usually occurs as beautiful, perfect tabular crystals. Excellent samples have come from the mines of Friedrichsegen, Dermbach and Schneeberg (E. Germany), Pribram and Mies (Czechoslovakia), Cornwall and Cumbria (England), Scotland, the Urals, Mexico, the USA, Australia and Italy (Sardinia, Valsugana and Monte Falo).

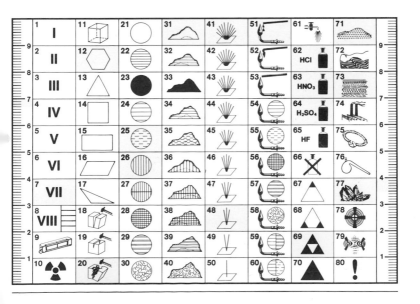

$(Fe'',Mg,Mn)_3(PO_4)_2.4H_2O$ Idaho (USA)

Iron magnesium manganese tetrahydrated phosphate. A rare mineral occurring as tabular crystals or granular aggregates; apple green with vivid vitreous lustre. It was first found in the Wheal Jane Mine at Truro (Cornwall, England); later it appeared at Stosgen near Linz (W. Germany) and in the pegmatites of Hagendorf (Bavaria). In the USA it is present due to the alteration of triplite in the pegmatites of New Hampshire, and as excellent crystals at Salmon (Idaho). It is of interest only to scientists and collectors.

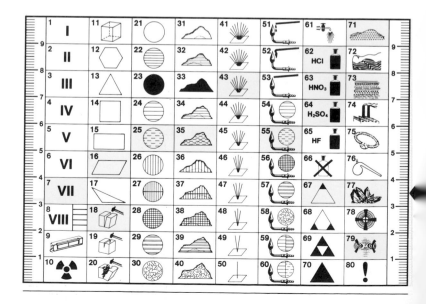

$Fe_3''(PO_4)_2.8H_2O$ 					Bolivia

Hydrated iron phosphate. Prismatic crystals, occasionally longer than a metre, in radial aggregates or shapeless masses. Colourless or white while fresh, it quickly alters to greenish blue. It forms in the outcrops of many sulphide deposits, or due to the alteration of primary phosphates in the pegmatites, or again in alluvial sediments rich in iron and organic matter. Splendid crystals, in groups or singly, have been found at Anloua (Cameroon) and Richmond (Virginia). Transparent samples with fine lustre come from Trepca (Yugoslavia); it occurs in the pegmatites of Hagendorf (Bavaria), in Cornwall (England), Llallagua (Bolivia), Utah and Colorado.

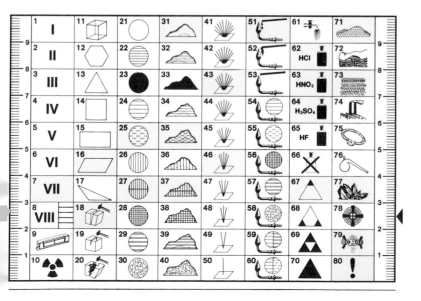

$Cu(UO_2|PO_4)_2.8–12H_2O$ Katanga (Zaire)

Hydrated copper uranium phosphate. It partly dehydrates when exposed to the air,
spontaneously producing metatorbernite which contains only eight molecules of
crystallization water. It is one of the most widespread uranium ores. It is also called
'green uranium mica' because of its vivid emerald-green colour. Precious samples with
tabular or flaky crystals come from Shinkolobve (Zaire), Joachimsthal and Saxony (E.
Germany), Tincroft (Cornwall, England), Mount Painter (Australia), the USA, and,
associated with autunite, parts of northern Italy.

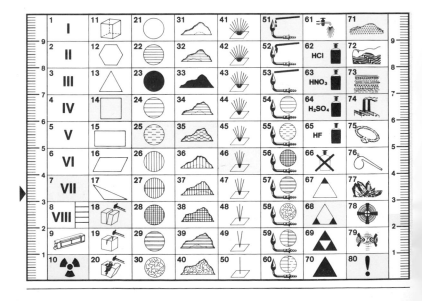

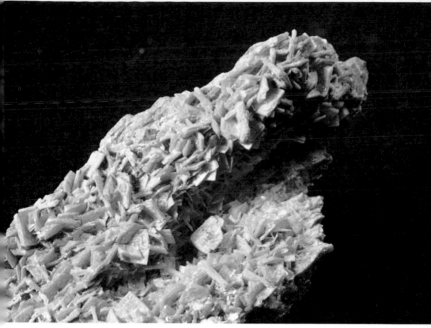

$Ca(UO_2|PO_4)_2.10-12H_2O$ Pcvcragno (Italy)

Hydrated calcium uranium phosphate. It occurs in flaky, tabular crystals greenish-yellow in colour. Calcium can be substituted, in small parts, by magnesium, vanadium, barium and lead. When exposed to heat, it loses water and changes to meta-autunite. The deposits at S. Symphorien near Autun (France), at Sabugal (Portugal), Johanngeorgenstadt and Falkenstein (E. Germany), Wolsendorf (Bavaria), Malagasy Republic and southern Australia are all well known. It can also be found in many parts of the USA: in the pegmatites of Bedford (New York), Keystone (Utah), Spruce Pine and Penland (North Carolina) and above all Mt Spokane (Washington). It also occurs in the pegmatites of Piedmont and those around Como (Italy).

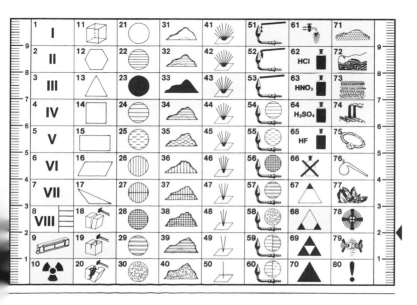

$(Mg,Fe)_2(SiO_4)$ Canary Islands

Silicate of magnesium and iron. Also called peridot. It is a solid, isomorphic solution of two minerals: forsterite $Mg_2(SiO_4)$ and fayalite $Fe_2(SiO_4)$. Olivines rich in magnesium but with a low iron content are used to produce low heat conductors and to extract magnesium. The transparent varieties, such as chrysolite, are highly valued as gemstones. The alteration of olivine produces many other minerals, such as serpentine, iddingsite, magnetite and opal. Splendid crystal samples come from St John's Island in the Red Sea, from Minas Gerais (Brazil), the Urals, Queensland (Australia) and the Faroe Islands.

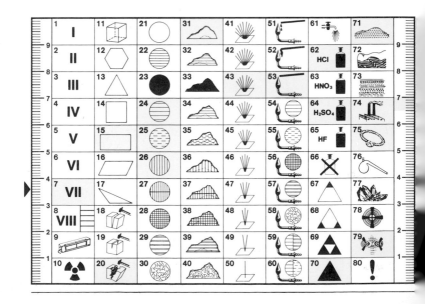

$$Ca_{10}(Mg,Fe)_2Al_4[(OH)_4(SiO_4)_5(Si_2O_7)_2]$$ Val di Susa (Italy)

Basic calcium aluminium silicate with iron and magnesium. First found on Vesuvius, hence its name; also called idocrase. It occurs in very beautiful varieties: the blue cyprine (Arendal, Norway), the fine, yellow xanthite (Amity, USA), the pale green wiluite (Siberia) and californite or 'American jade' (California and Oregon). Beautiful specimens come from Mexico, Canada, Switzerland and the Italian Alps: Val d'Aosta, Val d'Ala (Piedmont), Val Malenco and Val Camonica (Lombardy), Val di Fassa (Trento) and Val di Gava (Liguria). It is also found around Vesuvius and in Sardinia.

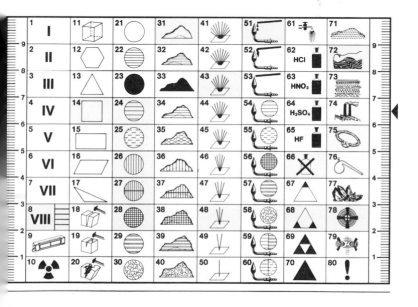

$Al_2Be_3(Si_6O_{18})$ Columbia (USA)

Beryllium aluminium silicate. Usually forms large hexagonal crystals and occurs in many varieties. An exceptional crystal 9m high and weighing more than 20 tons was found in Dakota. The deep green variety is the emerald (Columbia); the red one is morganite (Malagasy Republic, Brazil, Elba); the greenish-blue one, the colour of the sea, is the aquamarine (Brazil, the Urals, Malagasy Republic, Ireland, the USA); heliodor is greenish-yellow (South Africa, Malagasy Republic, Brazil); bazzite is the extremely rare blue variety rich in scandium (granite veins of Baveno (Italy), Switzerland, and the USSR). In Italy beryl is found in the pegmatites of the northern regions and in the granite of Elba.

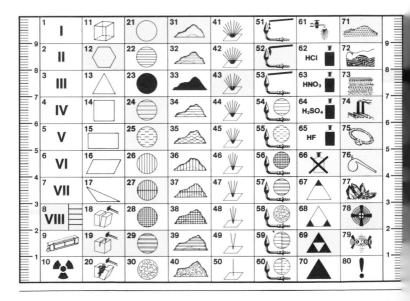

$Na(Li,Al)_3Al_6(BO_3)_3Si_6O_{18}(OH)_4$

Brazil

Complex borosilicate of aluminium and various metals. The formula indicates a series of minerals in isomorphic compounds. It is highly piezo-electric and pyro-electric. The varieties containing lithium are infusible. Very beautiful varieties occur, depending on the chemical composition: the colourless achroite; the brown dravite, rich in magnesium; pink elbaite and red rubellite, containing lithium; schorl, black, and blue indicolite, with an iron constituent; Brazilian emerald, where the green colour is produced by chromium. Beautifully clear tourmalines come from Elba (Italy), Minas Gerais (Brazil), the Urals, Mozambique, Malagasy Republic, Namibia and California. Opaque tourmalines have been found in the pegmatites of the Alps (Novara, Como, Valtellina and Val di Vizze).

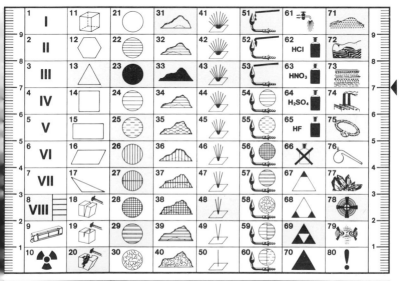

$Cu_6(Si_6O_{18}).6H_2O$　　　　　　　　　　　　　　　　Tsumeb (Namibia)

Hydrated copper silicate. A rare and splendid mineral, occurring as short prismatic crystals ending with rhombohedral faces. Exposed to heat, it dehydrates and blackens but will not melt. It forms in the oxidation zone of copper deposits. It has been found in calcite veins in the Kirghisi steppes, at Mt Altyn-Tube (Kazakhstan, USSR), at Mindouli and Katanga (Zaire), Copiapo (Chile), in Peru, at Atacoma (Argentina), in Arizona (USA), and above all at Tsumeb. It is used in the jewellery industry because of its vivid emerald-green colour, but it flakes easily.

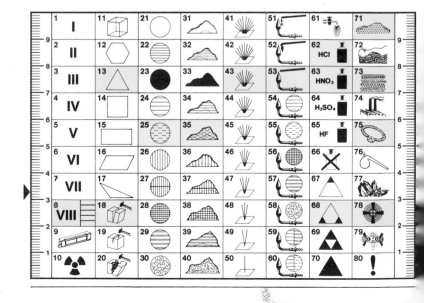

$KCa_2AlBe_2(Si_{12}O_{30}).\tfrac{1}{2}H_2O$ Switzerland

Hydrated potassium calcium beryllium and aluminium silicate. Its structure contains double hexagonal rings; its crystals are prismatic, elongated, pale green, white or colourless. Typical of the minerals of hydrothermal origin within alpine fissures. Classic deposits have been found in the Grisons (Giuf and Striem valleys, Switzerland) where the samples show crystals up to 3cm long. Recently it has been found in the Austrian Alps. Perfectly clear crystals come from Namibia and the Kola peninsula (USSR).

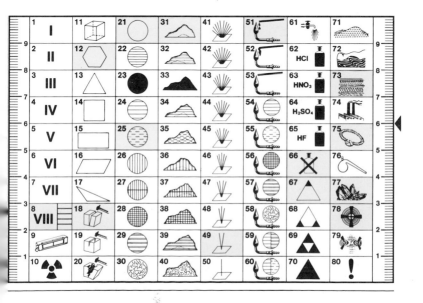

$CaFe(Si_2O_6)$ Campiglia Marittima (Italy)

Silicate of calcium and bivalent iron. It is the richest in iron of the minerals belonging to the isomorphic series of diopside. It is typical of the contact-metamorphic limestones. It is greenish black and rarely occurs as tabular crystals, more usually as radial, fascicular aggregates or compact, spathose masses. Its interest is purely scientific. Hedenbergite can be found at Capo Calamita (Elba), in the contact area between magnetite and limestone, at Campiglia Marittima, at Arendal (Norway), Obira (Japan), the USSR, Nigeria and Australia. It was given its name in honour of the Swedish scientist L. Hedenberg.

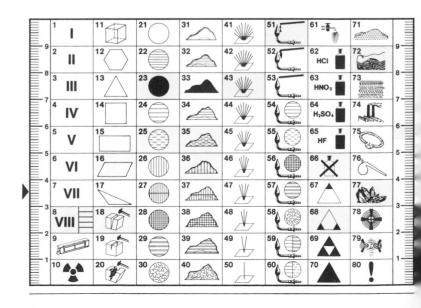

$$Ca_2(Mg,Fe'')_5[(OH,F)Si_4O_{11}]_2$$ Val d'Aosta (Italy)

Calcium magnesium iron silicate. It is the most important member in the isomorphic series of monoclinic amphiboles. It can contain manganese, thus forming the variety 'manganoactinolite'. Another variety is nephrite, a grey-green, compact aggregate which is used for jewellery (also called amphibolite jade). Very beautiful nephrites come from Central Asia, New Zealand and the USSR. They are often mistaken for jadeite, which is more precious. In Italy it has been found in Liguria and Basilicata. Actinolite is dark green, sometimes black and is very widely spread: excellent crystals are to be found in Austria (Zillertal) and Italy (Val Malenco, Val Germanasca).

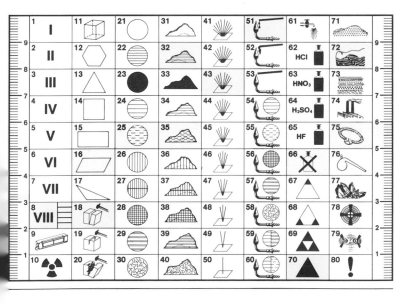

$(Mg,Fe)_7(OH|Si_4O_{11})_2$ Mera Valley (Italy)

Basic silicate of magnesium and bivalent iron. It is the most common among the rhombic amphiboles, but is not very widespread. Usually to be found as fibrous aggregates, rarely as distinct prismatic crystals, within crystalline schists, serpentines and peridotites; probably formed by the hydration of olivine. Valuable samples come from Kongsberg (Norway), Gothaab Fjord (Greenland), Onjarvi (Finland) and in the granites and mica schists of northern Italy. It has purely scientific interest.

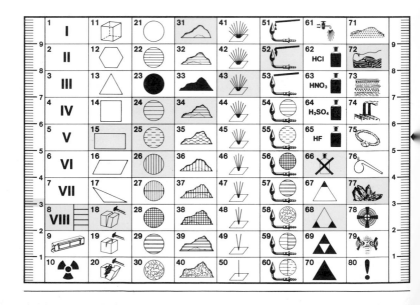

Apophyllite (c. × 1.5)

$KCa_4[F(Si_4O_{10})_2].8H_2O$ Poona (India)

Hydrated calcium potassium fluorosilicate. Its crystals can be bipyramidal, tabular, or lamellar; colourless, white, yellowish, or green. It is a secondary mineral formed by hydrothermal deposit within basalt cavities in association with scolecite, laumontite, stilbite, calcite, prehnite and analcime. It flakes with heating, hence its name (g. *apo* = detached; *fullon* = leaf). Excellent samples have been found at Poona, Kongsberg (Norway), Andreasberg (W. Germany), Nova Scotia (Canada), the Faroe Islands, Rio Grande do Sul (Brazil) and Alpe di Siusi (Italy). Of interest only to scientists and collectors.

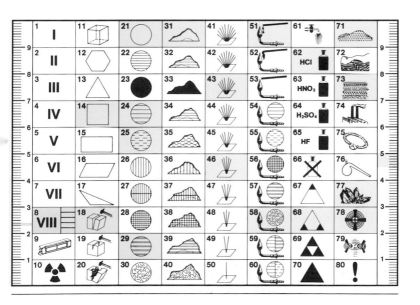

$Ca_2(Si_4O_{10}).4H_2O$ Poona (India)

Hydrated silicate of calcium. It belongs to the group of reyerite and zeophyllite. It is a rare mineral, usually found as pearly flakes joined together in small spherical or radial structures, or even as compact masses. It is found in basalt rocks in association with stilbite, laumontite, heulandite and apophyllite. Excellent samples have come from Poona, but it is also mined in other localities: Treshinish (Iceland), the Faroe Islands and Greenland. At New Amalden (California) it is found in association with apophyllite, in fibrous strata formed between apophyllite and the wall of the vein. It is also found in Nova Scotia (Canada), where it covers apophyllite.

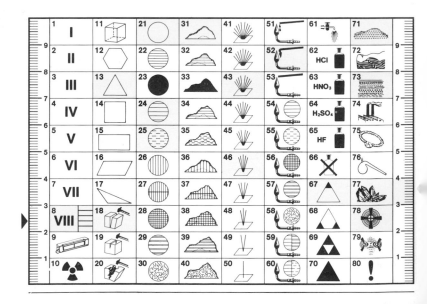

C   Minas Gerais (Brazil)

Pure carbon. Its crystals assume various cubic forms (octahedron, esacisoctahedron) with curved edges, and are usually colourless, yellowish, green, brown or black. The latter form the carbonado variety; others, radially fibrous, the bort variety. Diamonds have a hardness of 10.0, the hardest of all minerals. They are found in the kimberlite which fills the diatremes (volcanic vents) of South Africa, or in alluvial deposits. High levels of production have been achieved first in India, then in Brazil; today the majority comes from Africa. Top producers are Zaire-Kinshasa, Luderitzland, Ghana and Sierra Leone. Other diamonds come from the USSR, Borneo, Venezuela and Australia. It is indispensable for its use both in jewellery and industry in general (cutting and perforating).

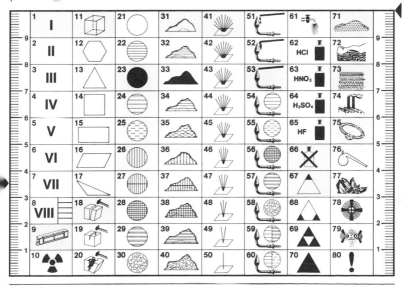

$Ca_3[(Al(F,OH,H_2O)_6)_2|SO_4]$ Colquiri (Bolivia)

Aluminium calcium hydrated fluoric sulphate. An extremely rare mineral occurring as monoclinic prismatic crystals, or acicular crystals, in radial aggregates similar in structure to wavellite. It can also be found as white, compact, granular masses. It is fragile, usually colourless or white, rarely reddish. It is formed by hydrothermal deposits in barite and fluorite veins, such as those of the Creede quadrangle, Mineral County (Colorado), together with gerarksutite, and in the oxidation area of the auriferous veins of quartz such as those at Tonopah and Nye County (Nevada). It has also been found at Colquiri. It is of interest only to scientists and collectors.

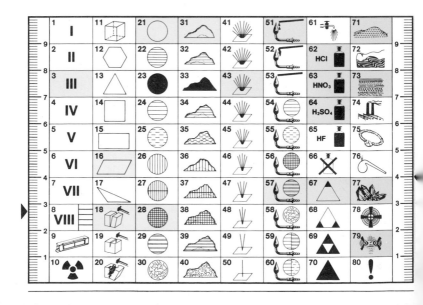

$Na_5(Al_3F_{14})$ Ivigtut (Greenland)

Sodium aluminium fluoride. The name derives from the fact that it looks like snow. Crystals are rare; it usually occurs as spathic, colourless or white masses. First found at Miask (Urals), it has since been discovered in great quantities at Ivigtut in association with criolite. The two are distinguished by the fact that chiolite crystals have perfect cleavage, unlike those of criolite. It is formed within pegmatitic rocks from solutions rich in fluorine. It produces hydrofluoric acid in contact with sulphuric acid. Together with criolite, it was used in the electrolytic process leading to the extraction of aluminium. It is now used for special glass and ceramic production.

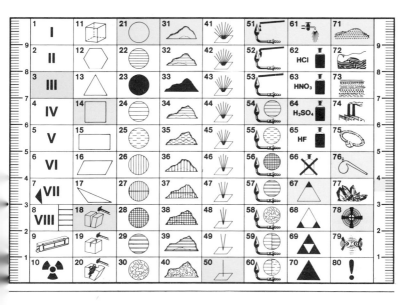

α-SiO₂

Brosso (Italy)

Silicon dioxide. The crystals, sometimes of gigantic proportions (up to 4m high), are usually hexagonal prisms terminating in two rhombohedrons which simulate an hexagonal bipyramid. Common quartz is the form alpha, stable up to 573°C; the hexagonal form beta is stable at higher temperatures. Polymorphic forms of quartz are cristobalite (tetragonal alpha and cubic beta), tridymite (hexagonal) and melanoflogite (cubic). Quartz is strongly peizo- and pyro-electric, and has a high rotating polarization. It crystallizes within the magma from geyserite and from the siliceous skeleton of radiolars and diatoms.

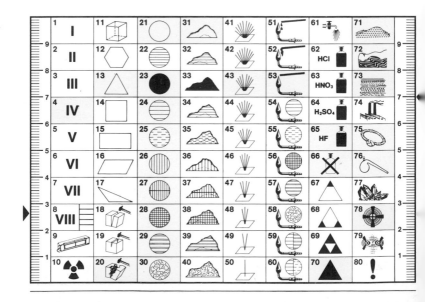

α-SiO₂ ... $\alpha\text{-}SiO_2$

Monte Ajù (Italy)

Silicon dioxide. When quartz contains trapezohedral faces it produces two types of crystals, not superimposable and one being the mirror image of the other (like two hands, left and right). It therefore produces left- and right-hand crystals which associate to form twins. If two right-, or two left-hand, crystals are intergrowing at an angle of 180°, a twinning is formed according to the 'Dauphinè law'; if, on the other hand, the two crystals are one left- and one right-handed the twinning occurs according to the 'Brazil law'. Rare twinnings are formed according to the 'Japan law', where two crystals intergrow flattened at right angles and form a heart shape, called 'heart-shaped twinning'.

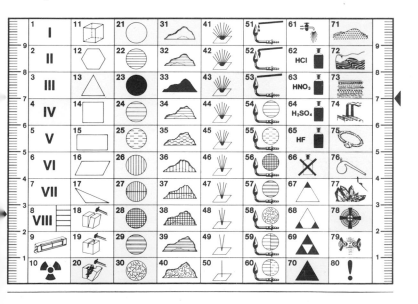

$CaMg(CO_3)_2$ Asturias (Spain)

Calcium magnesium carbonate. Very widely spread. Its origin is still uncertain; it may have formed as a consequence of the action of sea water on calcareous muds and organic deposits. It is the basic mineral of dolomia, a rock found in large quantities in the lithosphere. Whereas dolomite as such is not used, dolomia is used as building material, to make cement and refractory aggregates, and to extract magnesium. Dolomite is soluble in strong acids when these are heated. Splendid crystals have been found at Brosso (Italy), Binnthal (Switzerland), in the Asturias and Navarre (Spain), in Cornwall (England), and at Freiberg and Schneeberg (E. Germany).

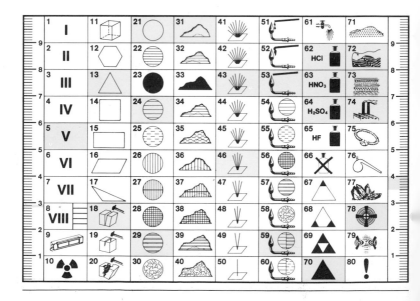

$Ca[B_3O_4(OH)_3].H_2O$ Bigadic (Turkey)

Hydrated calcium borate. It occurs as clear crystals, usually equidimensional, colourless, white and many-faceted. It can be mistaken for calcite or datolite. It is a very important mineral for use in the extraction of boron. It is soluble when heated in hydrochloric acid; the cooling down process produces boric acid. The largest deposits are found in California (Death Valley), Turkey (Bigadic), Chili, Argentina and Kazakhstan (USSR), Boron is included in alloys, used as fuel for missiles, and its salts are utilized in chemistry, the pharmaceutical industry, cosmetics, enamels, steel and glass.

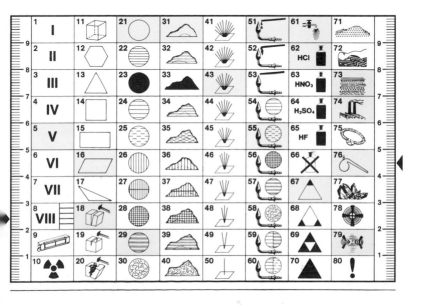

$Ca(SO_4).2H_2O$ Zapala (Argentina)

Hydrated calcium sulphate. Occurring as clear, colourless crystals, occasionally white, yellow or brown, gypsum often forms twins in a 'swallow-tail' or 'spear-head' shape. It can form compact zonal masses (chalky alabaster), fibrous crystals (sericolite), elongated, transparent, laminar crystals (selenite) or furrowed, lamellar masses (desert roses). Gypsum deposits are distributed worldwide. The most noted are near Volterra (Tuscany; called alabaster), in Sicily (in association with sulphur and celestine), Spain, Egypt, France, England (Matlock, Swanage, Isle of Sheppey), Poland and Czechoslovakia.

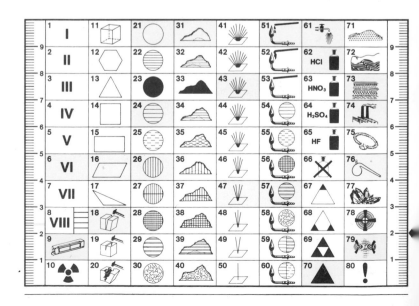

$Zn_4[(OH)_2|Si_2O_7].H_2O$ Mexico

Basic hydrated zinc silicate. The name is derived from the shape of its crystals, which differs at the two extremities of the vertical axis. Crystals are usually small and infrequent. It is considerably piezo- and pyro-electric: crystals, when heated or submitted to pressure, acquire different electric charges at the two extremities. It is found worldwide in lead and zinc deposits: it has been mined in Westphalia and Rheinland (W. Germany), Bleiberg-Kreut (Austria), Cumbria (England); and Vieille Montagne (Belgium). Excellent crystals come from Chihuahua (Mexico), various localities in the USA (Leadville, Whyte County, Sterling Hill, Friedensville), in Bolivia, Peru and Australia.

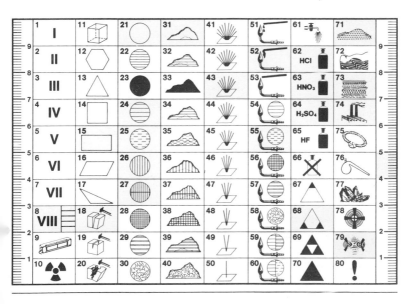

CaTi(O|SiO₄)   Val di Gava (Italy)

Calcium titanium silicate. Also called sphene after the wedge-like shape of its crystals. Its varieties are grothite (containing iron), greenovite (containing manganese) and yttriotitanite (containing yttrium and other elements). Splendid crystals are to be found in the Alpine lithoclasses: in Italy in Val di Nevero, Val d'Ala, Val d'Aosta, Valle del Sangone, Valle Germanasca (Piedmont), Val Malenco (Lombardy), Val di Vizze and Val Passiria (Alto Adige); in Switzerland (Zermatt, Tavetsch), the Urals, the Kola Peninsula (USSR), Mexico and the USA. When found in large masses, it is used for the extraction of titanium; if clear and beautifully coloured, it is used for jewellery.

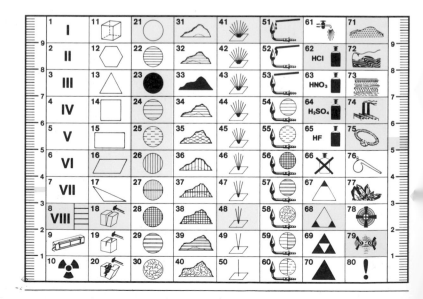

$CaMg(Si_2O_6)$ Pian della Mussa (Italy)

Magnesium calcium silicate. Prismatic, often elongated slender crystals; also fibrous radiating aggregates. One of a series of three isomorphs together with johannsenite and hedenbergite. It occurs in several varieties: violane, containing manganese; chromediopside, with chrome; jeffersonite, green, with zinc and manganese; coccolite, with iron. Sahlite (Sahla, Sweden) is the isomorphic amalgam of two parts diopside and one of hedenbergite; ferrosahlite is the opposite. Extremely beautiful samples of diopside have been found in Switzerland (Binnthal), Austria (Zillerthal), Sweden (Nordmarken), the Urals and Lake Baikal and Italy (Trento, Piedmont and the ejecta of Vesuvius).

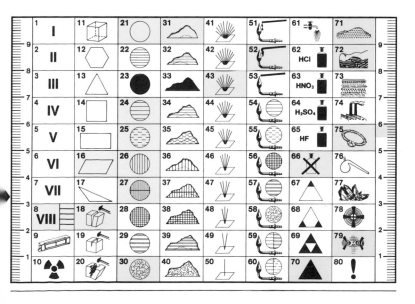

$KAl_2[(OH,F)_2AlSi_3O_{10}]$ Minas Gerais (Brazil)

Basic aluminosilicate of potassium and aluminium with fluoride. Sometimes called 'potassium mica' or 'white mica', it is the most common of the micas. It is monoclinic, and its tabular crystals have an hexagonal or trigonal contour. It is widely used in industry owing to its hardness and its electrical and thermal-insulating properties. Varieties are: alurgite, containing manganese; ferro-muscovite, with iron; and fuchsite, containing chrome. Huge tabular crystals, with surfaces up to 5sq m, have been found in Ontario, South Dakota, New Hampshire, Brazil and India. In Europe, it is mined in the Alps, although the scale of this variety is too small to be used industrially.

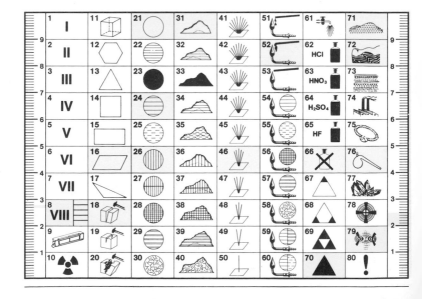

$Ca_8[(Cl_2, SO_4, CO_3)_2(Al_2Si_2O_8)_6]$

Mt Somma, Vesuvius (Italy)

Complex calcium aluminosilicate. One of the minerals in the scapolite series, which includes isomorphic amalgams in all proportions of two components: marialite and meionite. The first member of the series is marialite, which contains up to 20 per cent of meionite; the last member is meionite, which can in turn contain up to 20 per cent of marialite. Intermediate members are dipyre and mizzonite. Meionite occurs as vertically-striated prisms and fibrous aggregates. It is found, in its almost pure state, on the island of Elba and in the calcareous ejecta of Vesuvius (Mt Somma); beautiful, though less pure, crystals have been found in Switzerland and the Apennines.

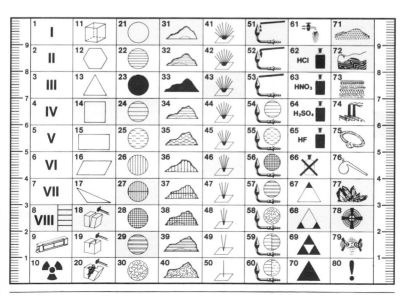

$Ba(Al_2Si_6O_{16}).6H_2O$ Kersnas (Finland)

Hydrated barium aluminium silicate. One of the zeolites. Has pseudo-rhombic crystals which are often twinned in a cross. It breaks under heat, is resistant to fusion and its barium content imparts a yellow-green colour to a flame. It is typical of basalts, phonolithic and trachitic. Notable samples have been found at Strontian (Scotland), Andreasberg (W. Germany), Rudolstadt (E. Germany), Kongsberg (Norway) and Kersnas (Finland). In Canada it has been found on Mount Rabbit, on the northern coast of Lake Superior; and, in the USA, within the gneiss deposits underneath New York.

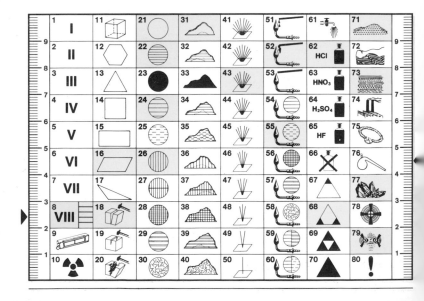

$(Ca,Na_2)(Al_2Si_4O_{12}).6H_2O$ Gadd's Hill (Tasmania)

Hydrated calcium, sodium and aluminium silicate. Its idiomorphic crystals are rhombohedral, pseudocubic and often twinned through cross-like intergrowth. Chabazite is a zeolite frequently found in basalts. Splendid crystals come from the basalts of Ireland, the Faroe Islands and Skye; others from Aussig (Czechoslovakia), Oberstein (W. Germany), Richmond (near Melbourne, Australia), New Jersey and Oregon, the Alps north of Bolzano and the basalts near Vicenza (Italy).

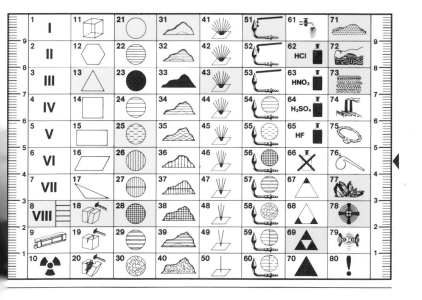

CaF$_2$ Weardale, Co Durham (England)

Calcium fluoride. Very widespread as perfect monometric crystals, usually cubic, rarely octahedral or in other monometric shapes. Its colour is extremely variable due to inclusions or faults in its structure. Most of the samples are fluorescent, violet-coloured, when exposed to Wood's filter lamp. This phenomenon, fluorescence, is so called after fluorite. In Europe, the famous violet, green and white fluorites can be found in Cumbria and Derbyshire (England); the yellow ones come from Wolsenberg (Germany); the pink ones from St Gotthard (Switzerland); the green and violet ones from Spain; the blue ones from France; the green ones from Norway and the USSR, and the clear, colourless ones from Corvara (Bolzano, Italy) and other alpine localities.

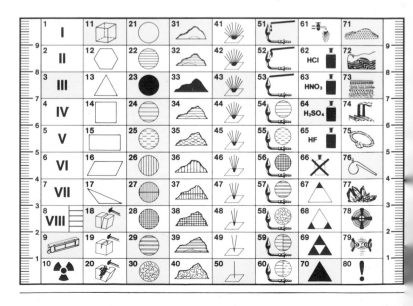

CaCO₃   Morocco

Calcium carbonate. The sample above derives its beautiful violet-pink colour from traces of cobalt which substitutes calcium. Calcite is a very widely spread mineral. It is very clear or variously coloured, with beautiful, rhombohedral or scalenohedral, often twinned, crystals. It can contain various cations and gives origin to several varieties: manganocalcite, ferrocalcite, strontium calcite, zinc calcite, magnesium calcite and cobalt calcite. The latter should not be confused with the homonimous species (CoCO₃), which is rare, occurs as minute crystals and is also called spherocobaltite. Calcite can also be found as compact, microcrystalline masses (limestones), saccaroids (marbles), zoned (alabasters) and as curious compound formations: stalactites (with a centre groove), stalagmites and coral-like or mamillary aggregates.

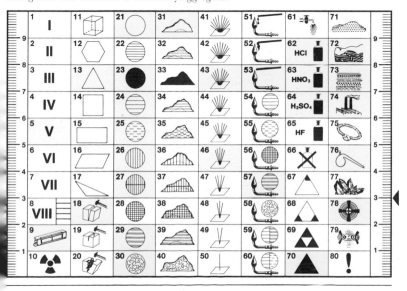

$Ca_2Mn_7(Si_5O_{14}OH)_2.5H_2O$ California (USA)

Hydrated silicate of calcium and magnesium. Also called rhodotilite. It rarely occurs as small prismatic crystals; it is more usually found as fibrous, radiating masses or small spheres. It belongs to the rhodonite series. Its colour varies from pinkish-violet to pink, brownish-yellow and golden yellow. Often found in association with manganese deposits, particularly in the contact area with base rocks. It is often associated with calcite. It has been found in the manganese deposits at Nanzenbach (near Dillenburg, W. Germany), Jakobsberg and Langban (Sweden); in the gold veins of Idzu (Japan); and at Broken Hill (NSW, Australia). Splendid samples have come from the Hale creek mines (Trinity County, California). Can be corroded by acids, under heat.

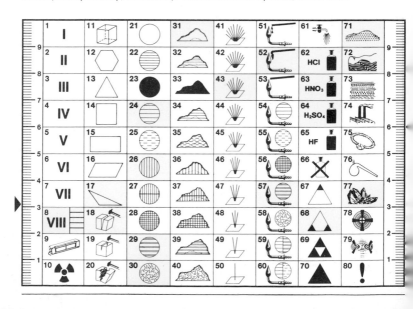

HgS   Almaden (Spain)

Mercuric sulphide. Rarely as well-formed rhombohedral crystals, usually as granular masses or earthy coatings, sometimes as prismatic needles. Once used as a colourant because of its vivid, vermilion-red colour. Does not melt with heat, it sublimates at 580°C and condenses as mercury. The principal mines, in which cinnabar, together with other sulphides (antimonite, realgar), impregnates magmatic rocks or limestones, are: Almaden (Spain), Abbadia San Salvatore (Mt Amiata, Italy) and Idrija (Yugoslavia). Further deposits have been found in Hunan (China), Nikotowka (USSR), Algeria, Huancavelica (Peru) and Pike County (Arkansas).

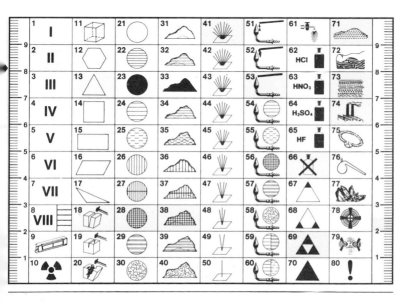

Sb_2S_2O Stabiello (Italy)

Antimony oxisulphide. The name is derived from a Persian word meaning 'bright red', and thus it was named by the alchemists of old. It occurs as tufts of acicular crystals, cherry-red, with adamantine lustre; also as earthy coatings. It originates from the alteration of antimonite and can be found in the latter's deposits. It is often associated with senarmontite, valentinite, cervantite and stibiconite. It has been found in E. Germany (Freiberg), Czechoslovakia (Perneek and Pribram), Canada (Quebec and Nova Scotia), the USA (California and Idaho) and Italy (Sardinia and Tuscany).

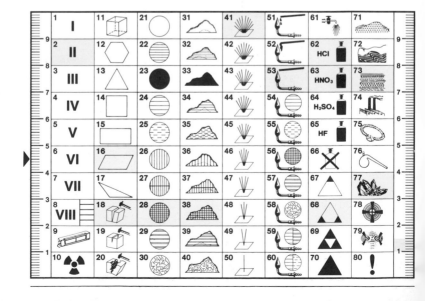

Cu₂O   Tsumeb (Namibia)

Cuprous oxide. Its crystals are cubes, octahedrons, rhombododecahedrons; occasionally the cubes elongate into acicular crystals, thus originating the chalcotrichite variety. Its colour is ruby red, altering to green when exposed to the air, due to the formation of malachite. Well known are the malachite pseudomorphoses forming on cuprite rhombododecahedrons which have been found in the famous deposits of Chessy (Lyon, France). Cuprite is found in copper deposits in association with malachite and azurite. Splendid crystals have been found in the mines of Redruth and Liskeard (Cornwall, England), in the Urals, in the Altai, Tsumeb (Namibia), Bisbee and Morenci (Arizona), Corocoro (Bolivia) and Chuquicamata (Chile).

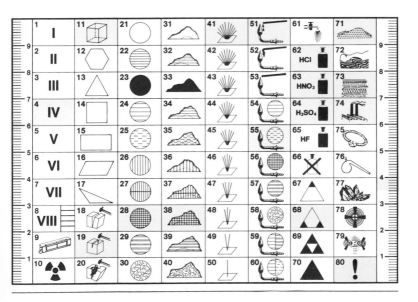

Al$_2$O$_3$ India

Aluminium oxide. Barrel-shaped pseudo-hexagonal crystals, sometimes tubular; also brown granular masses (emery). Due to its hardness (9° on the Mohs scale) it is used as an abrasive. When it is clear and occurs in splendid colours, it provides extremely valuable gems: ruby, when red; sapphire if blue; oriental topaz, when yellow; oriental amethyst when violet; and oriental emerald when green. The most famous source of rubies is Modok (Burma); of sapphires, Thailand, Campuchia, India and Australia (Queensland). Variously coloured crystals can be found in Sri-Lanka and Malagasy Republic. Large deposits of emery have been found in Smyrne (Turkey), Naxos and Samos (Greece) and in the Urals.

γ-TiO$_2$ Lavagna (Genoa, Italy)

Titanium dioxide. A member, with rutile and anatase, of the series of polymorphic modifications of titanium oxide. It occurs as beautiful, rhombic crystals, full of lustre, brownish-red, grey or black, usually tabular, often vertically striated. It can be found in the lithoclases of alpine schists, often in association with anatase. The best samples, up to 5cm long, have come from the area of Maderaner Thal (Grisons, Switzerland) and from Intschi (St Gotthard, Switzerland). Fairly good samples have come from Bourg d'Oisans (France), Tête Noire (Mont Blanc), Abilch and Pragratten (Austria) and Sondalo and Beura (Italy). It is also present in the Urals and Arkansas.

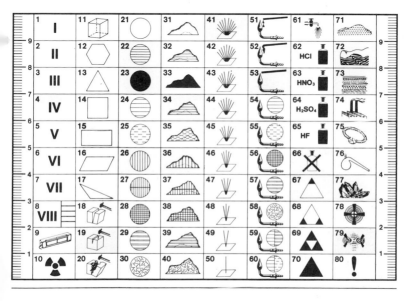

MnCO$_3$ Silverton, Colorado (USA)

Manganese carbonate. It belongs to the isomorphic series of calcite. It occurs as rhombohedral, scalenohedral crystals, or, more frequently, as stalactitic, mamillary or granular masses. The oxidation of manganese turns it from pink to brown. When heated, it is soluble in hydrochloric acid. It is used for the manufacture of ornamental objects and, if mined in quantity, for the extraction of manganese. The largest crystals, up to 5cm long, come from Colorado (USA), Rumania, Freiberg (E. Germany), and Peru. Brilliant scalenohedral crystals come from Kuruman (South Africa) and Gabon (Africa). Large deposits have been found at Huelva (Spain), Ariege (France) and Tschiaturi (USSR). The most beautiful zoned rhodochrosite came from Catamarca (Argentina). It is also found in Val d'Aosta, Liguria and Sardinia (Italy).

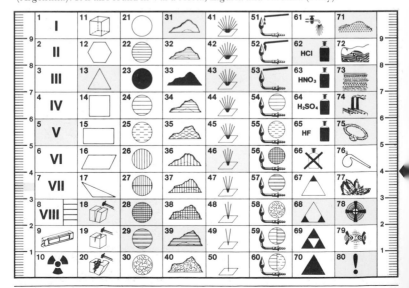

$Pb(MoO_4)$ Hungary

Lead molybdate. It occurs as splendid crystals, whether tetragonal, tabular or scaly, occasionally pseudo-octahedral or pseudo-cubic; also as granular, earthy or scaly masses. It is a secondary mineral of lead deposits together with cerussite, pyromorphite, vanadinite and mimetite. Only when found in large quantities can it be used for the extraction of molybdenum. The honey-yellow crystals of Bleiber (Austria) and Mesica (Yugoslavia) are famous; so are the red crystals of Rezbanya (Rumania), Congo, and Arizona (Red Cloud); and the colourless ones of Tsumeb (Namibia) and Phoenixville (Pennsylvania). Incredible samples have come from Chihuahua (Mexico), Morocco, Australia, and some American mines. Other European deposits can be found in Sardinia and in some of the valleys of northern Italy.

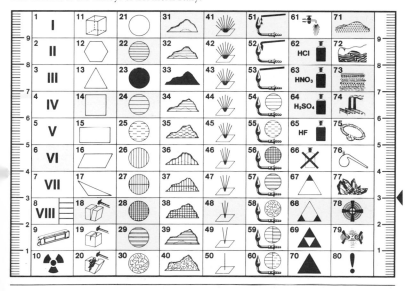

$Pb_5[Cl(VO_4)_3]$ Mibladin (Morocco)

Lead chlorovanadate. A secondary mineral in lead deposits like wulfenite. It occurs as fine hexagonal prismatic crystals, varying from orange-red to bright red, sometimes to brown. Also as crusts or fibrous masses. Vanadium is extracted from it and used in metallic alloys and chemical processes. Valuable deposits have been found in the mines of Old Yuma (Arizona), New Mexico (Hillsboro and Lake Valley), Beresovsk (Urals) and Kärnten (Austria). At Grootfontein (S. Africa) 5cm long crystals have been discovered, and Oudjda (Morocco) has yielded excellent and abundant samples. Also valuable are the vanadinites of Cordoba (Argentina), and Los Lamentos (Mexico).

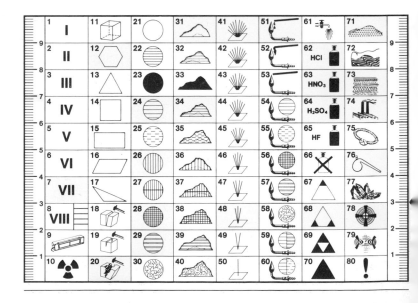

$Ca_3Al_2(SiO_4)_3$　　　　　　　　　　　　　　　　Val di Susa (Italy)

Calcium aluminium silicate. A typical garnet in which the calcium aluminium silicate prevails and can be partially substituted by bivalent iron. It crystallizes into rhombodecahedrons and icosatetrahedrons which can be colourless, white, yellow-orange (hessonite), green, pink or red. Garnets are excellent gemstones from the point of view of hardness, colour and lustre; they are the jewels of the Alps. Splendid samples have come from Val d'Aosta, Val d'Ala, Valle del Sangone and Val di Susa (all in Piedmont, Italy). Also famous are the garnets from Asbestos (Canada), Ramona (California) and Malagasy Republic.

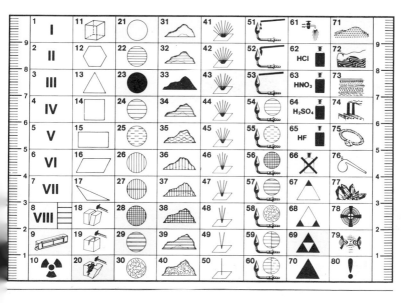

$CaMn_4(Si_5O_{15})$ Broken Hill (Australia)

Calcium manganese silicate. It occurs as large, triclinic, tabular crystals, or as compact, granular masses. It often shows darker veins or patches due to the oxidation of manganese. Its colour can be pink, deep red or brownish-red. It can contain iron, magnesium and zinc: hsihutsunite is a variety which contains magnesium; fowlerite contains zinc. Crystals are used in jewellery. Excellent crystallized samples come from Broken Hill (Australia), Chikla (India), Pajsberg and Langban (Sweden), Simsio (Finland), the Arrow Valley (New Zealand), Sverdlovsk (Urals), Brazil, South Africa, and Val d'Aosta (Italy).

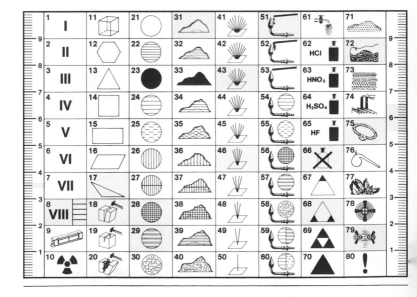

$(Mg,Cr_{<3})[(OH)_2AlSi_3O_{10}]Mg_3(OH)_6$ Turkey

Basic magnesium chromium aluminosilicate. A chromium-rich chlorite formed by the alteration of olivinic rocks, rich in chromium, into serpentines. It is monoclinic and crystallizes as pseudohexagonal scales: deep violet, pinkish violet, reddish or even greenish-blue. Magnesium can be partly substituted by bivalent iron. It is found, as small scales, within the 'Iherzolite' (peridotitic, olivino-pyroxenic rock) of Piedmont (Italy)—where the largest scales are up to 2 × 4cm—as well as in the serpentines of Val Malenco and Val d'Aosta (Italy), at Kraubath (Austria), Miass (Urals) and Texas (USA).

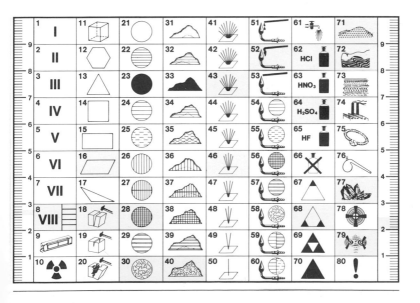

$Ca(Al_2Si_7O_{18}).6H_2O$ Val di Fassa (Italy)

Hydrated calcium aluminium silicate. A zeolite, it produces idiomorphic, monoclinic crystals. It can be colourless, white, greenish or yellow-pink. Natronheulandite is a variety containing sodium; bariumhuelandite contains barium; clinoptilolite is very rich in silicon. Heulandite is a secondary mineral forming in basalts, andesites and diabases. Superb samples come from Poona (India), Ireland, the Faroe Islands, Nova Scotia (Canada), New Jersey and West Paterson (USA); beautiful crystals are also found in the basalts of the Alps and the andesites of Sardinia.

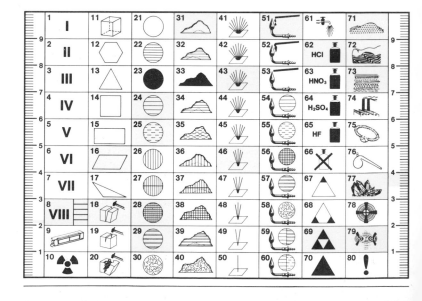

NaCl   Hessen (E. Germany)

Sodium chloride. It occurs as extremely clear cubes, which can also be white, pink or red, due to inclusions. It occasionally shows blue or violet areas due to faults in the crystals' lattice or to radiations, probably from radioactive potassium. Luminescent in ultraviolet rays. It is found in large deposits due to the evaporation of saline waters; the most important ones are those of Wieliczka (Poland), Stassfurt (E. Germany), Cardona (Spain), Dax, Vic and Dieuze (France), Salzkammergut (Austria), Armenia, Turkestan, Calabria, Sicily and Tuscany (Italy). Various deposits can also be found in the USA, well-known among them being Searles Lake, California, where a thick crust of rock-salt covers the water.

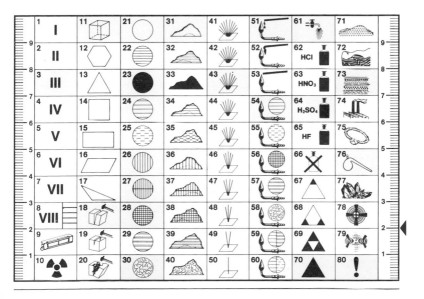

MgCO$_3$ Salzburg (Austria)

Magnesium carbonate. The most important magnesium ore. Rhombohedral crystals are rare; it usually occurs as earthy or fibrous crystalline masses. It can contain varying amounts of iron in a solid solution, which determines the various varieties: breunnerite, mesitine and pistomesite. It is due to the alteration of peridotitic rocks and serpentines. It can accompany, as gangue, metallic veins. Famous deposits are to be found in Austria (Styria), Sweden, Norway, the Urals, Madras, Korea, Manchuria, Canada, the USA (California and Nevada) and Italy (Piedmont and Tuscany).

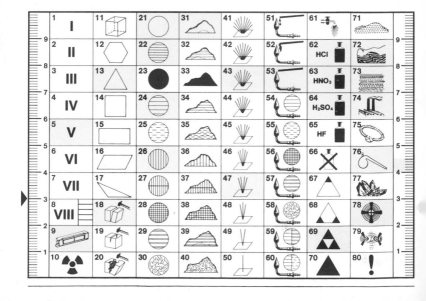

CaCO$_3$　　　　　　　　　　　　　　　　　　　　　　　　Missouri (USA)

Calcium carbonate. This particular sample is constituted by scalenohedrons with internal shades due to an interruption in the crystals' development. Calcite is the stable form of calcium carbonate (trigonal system); aragonite (rhombic) and vaterite (hexagonal) are the polymorphic modifications. Splendid calcite crystals exist worldwide. 'Iceland spar', a very clear calcite used in optics, is famous and comes from Helgustadir. Very beautiful are the calcites of Kapnik (Czechoslovakia), Kongsberg (Norway), Langban (Sweden), Cumbria (England); Fontainebleau (France), Kara Dag (Crimea), New Jersey, Missouri and Lake Superior (USA), Mexico, the Tuscan Apennines and Sardinia. Calcite is used in the manufacture of cement, in the metallurgic and building industries (marble), chemistry and physics.

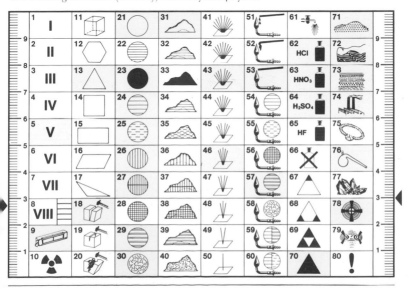

$CaFe''(CO_3)_2$ Traversella (Italy)

Calcium iron carbonate. A mineral analogous to dolomite, from which it differs in its iron rather than magnesium content. Also called 'brown spar'. It is widespread but only appears in small quantities. It is usually found in the gangue of metallic veins, or as a secondary deposit mineral in primary rocks, such as diorites and porphyry. Its crystals are rhombohedral, as in dolomite, also lentiform, often concave. Fine samples have come from the mines of Traversella (Turin), Cuasso al Monte and Cavagnano (Varese, Italy), Binnthal (Switzerland) and Bourg d'Oisans (France). Soluble only in heated acids. Of interest only to scientists and collectors.

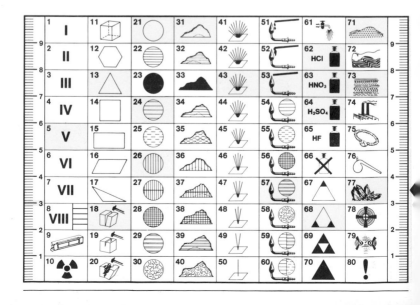

CaCO₃   Agrigento (Sicily)

Calcium carbonate. The name is derived from Aragon (Spain), the region where the mineral was first found. Its crystals are rhombic, prismatic, and often twinned three at a time, thus simulating hexagonal prisms. It also occurs as massive, concretional, stalactitic, fibrous-radial or coral-like (the variety 'flos ferri'). It is usually colourless, sometimes white, reddish, yellow, green and blue. It is a dimorph of calcite at high pressure and slowly reverts to calcite. Fine samples come from Molina de Aragon (Spain), Agrigento (Italy), Bastennes (France) and Alston Moor (Cumbria, England). The 'flos ferri' variety comes from Styria (Austria).

SrCO$_3$ Oberdorf (Austria)

Strontium carbonate. First found at Strontian (Argyllshire, Scotland), hence its name.
A member of the rhombic isomorphic series of aragonite. It rarely occurs as idiomorphic
crystals, more usually as granular, fibrous or columnar masses. It is colourless or white,
sometimes pink, grey or green. It was formed by the action of warm subterranean waters
of meteoric origin; it is found in some limestones and is associated with the gangue of
certain ore veins. Workable deposits are found in Westphalia and Harz (W. Germany),
Spain, Strontium Hills (California) and Mexico. Beautiful crystals come from Strontian
(Scotland), Leogang (Austria) and Braunsdorf (E. Germany).

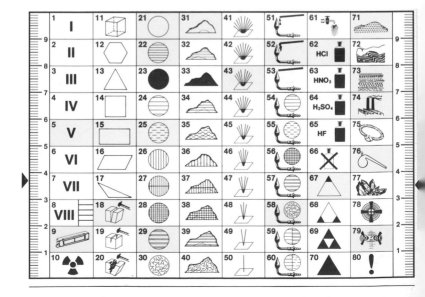

BaCO$_3$ Illinois (USA)

Barium carbonate. A member of the aragonite series. It occurs as rhombic crystals, often trillings, which simulate an hexagonal bipyramid, or as globular formations with parallel streaks, or as granular and fibrous masses, white or grey. It is of hydrothermal origin due to the reaction of barium-rich solutions with carbonated rocks. It can contain strontium and calcium, and is often accompanied by calcite and dolomite. Very beautiful crystals come from Hexham (Northumberland, England), Alston Moor (Cumbria, England), Karakala (USSR), Leogang (Austria), Tsumeb (Namibia), Lexington (Kentucky), El Portal (California), Thunder Bay (Ontario) and Sardinia.

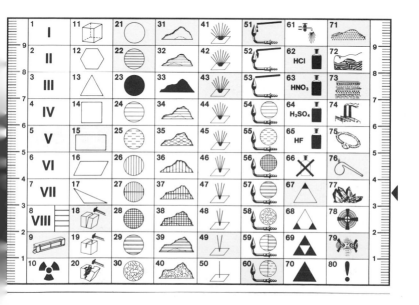

PbCO$_3$ Tsumeb (Namibia)

Lead carbonate. A member of the isomorphic series of rhombic carbonates, it occurs as compact masses; earthy, stalactitic concretions; or clear crystals, either tabular or elongated, often as multiple twins, pseudohexagonal and in radiate arrangements. It is colourless or white, sometimes yellowish or green due to inclusions. It is a secondary mineral of the oxidation zone of lead deposits with anglesite, smithsonite, pyromorphite, etc. It is found at Mies (Czechoslovakia), Rezbanya (Hungary), Ems (W. Germany), Andalusia, Murcia and Almeria (Spain), Laurium (Greece), Mindouli (Zaire), and Sardinia. The best crystals come from Tsumeb, Broken Hill (Australia) and Dona Ana (New Mexico).

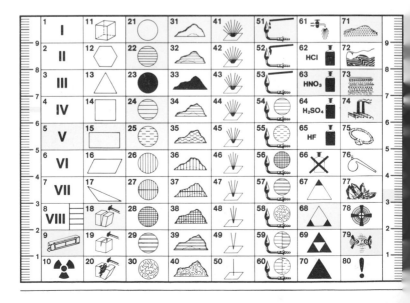

$Pb_2(Cl_2|CO_3)$ Monteponi, Sardinia

Lead chlorocarbonate. So called because it was believed to be a derivative of phosgene ($COCl_2$). It is a secondary mineral found in the oxidation zone of lead deposits. It may have formed under the action of sea waters, or of running waters rich in chlorates, on galena. The most famous samples, displayed in museums throughout the world, came from Monteponi, where a crystal was found to weigh 45kg. Crystals are short and tabular, usually very clear, and very much sought after by collectors. It is mined at Laurium (Greece), Tsumeb (Namibia), Tainowitz (Poland), Matlock (Derbyshire, England) and Custer County (Colorado).

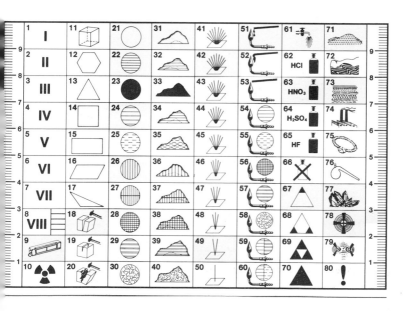

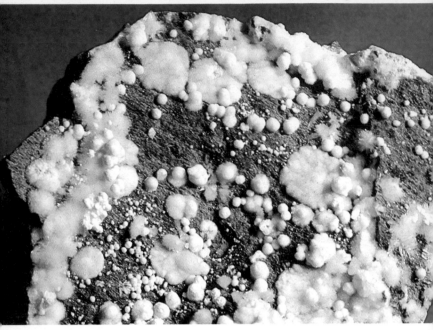

$Mg_5[OH(CO_3)_2]_2.4H_2O$ Montjovet, Val d'Aosta (Italy)

Hydrated magnesium carbonate-hydroxide. It occurs as minute, acicular or bladed white crystals, arranged as thin radial incrustations or powdery masses. It originates as a secondary hydrothermal mineral, at very low temperatures, in the fissures of rocks containing magnesium, such as the peridotites and serpentines; it can also form as a consequence of contact metamorphism between magmatic masses and dolomitic limestones. It is often associated with brucite, asbestos and aragonite. Of interest only to scientists and collectors. It is mined at Montjovet, Val di Viu and Val Malenco (Piedmont, Italy), in the dolomites of Monte Somma (Vesuvius) and in many serpentines all over the world. The picture shows it in association with aragonite, represented by the less white nodules.

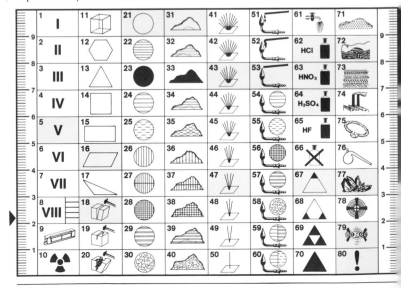

$$Mg_2[(OH)_2|CO_3].3H_2O$$ Montjovet, Val d'Aosta (Italy)

Basic hydrated magnesium carbonate. A rare mineral discovered in 1902 in the asbestos caves of Val Malenco (Sondrio, Italy). It occurs as small monoclinic, acicular crystals, usually grouped into radiated globes, small bunches, or fibrous-radiated crusts. It is greyish white in colour. It is formed by the alteration of serpentine rocks caused by carbonated waters at very low temperatures. It is often associated with brucite and asbestos, and is found in the serpentines of Val Malenco and other localities in Piedmont (Italy), Austria, Yugoslavia, Nevada and New Jersey (USA). Splendid crystals were found in the Val d'Aosta during the construction of a motorway.

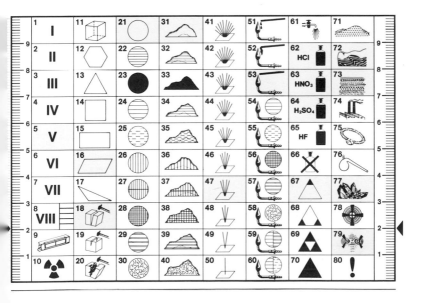

$Na_2[B_4O_5(OH)_4].8H_2O$ Boron (USA)

Hydrated sodium borate. It occurs in compact masses or short, prismatic crystals, often tabular and very beautiful. It has a sweetish, cloying taste. When exposed to the air, it partially dehydrates and turns to tincalconite. It is found in the deposits formed by the evaporation of boraciferous and salty waters of various lakes; in arid areas it can be found as efflorescences on the soil. Samples come from Kashmir, Tibet and Iran; economically important deposits exist in the USA (Borax Lake, Boron, Furnace Creek and Searles Lake), which have produced the best specimens with crystals often of very large dimensions.

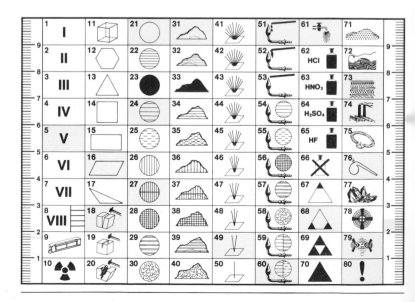

$Pb(SO_4)$ Monteponi, Sardinia

Lead sulphate. Its crystals are usually tabular or prismatic, rarely bipyramidal; it produces encrustations and granular, earthy or compact masses. Fine crystals have come from Muesen (W. Germany), Anglesey (Wales; hence its name), Spain and Tunisia; Italy is famous for the splendid prismatic or bipyramidal crystals, black, white or yellow, found at Monteponi, and for the much rarer ones of Montevecchio, which are green and transparent and grow on galena. But perhaps the best specimens come from Tsumeb (Namibia), Wheatley Mine (Arizona) and Los Lamentos (Mexico). Like the rest of lead ores, it is used for the extraction of that metal.

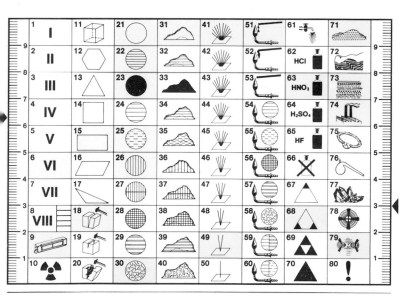

$K_2Mg(SO_4)_2.4H_2O$ Leopoldshall (W. Germany)

Hydrated potassium and magnesium sulphate. It occurs as monoclinic, elongated crystals, often as lamellar twins formed by two individuals set at 60°. It is colourless and transparent, occasionally yellowish, and with a bitter taste. It is of secondary origin, as part of the salt deposits left by the evaporation of sea waters. It hydrates when exposed to the air and turns to picromerite by absorbing two water molecules. It is mainly found at Westerregeln, Leopoldshall, Neu Stassfurt and in most of the potassium districts of Germany; also in New Mexico, in the innermost parts of salt deposits, in association with kainite, polihalite, halite and silvite. Good crystals have been found in the salt mines of Sicily.

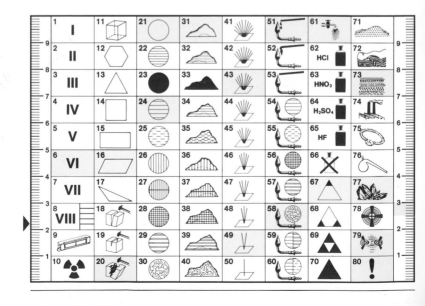

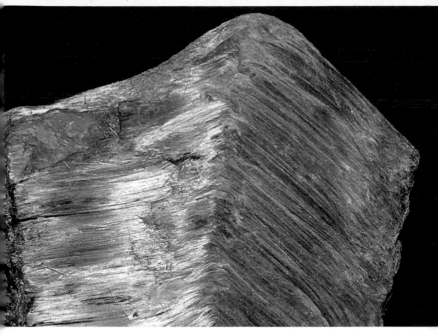

$Ca_2Mg_5[(OH,F)Si_4O_{11}]_2$ Val di Lanzo, Turin (Italy)

Hydrated calcium magnesium silicate. The sample presents a herringbone pattern. Tremolite can contain manganese, in which case it is called manganotremolite. It occurs as fibrous, columnar crystals, occasionally radiated (it is then called grammatite). It is widespread in dolomitic and calcareous rocks rich in silica, which have totally re-crystallized as a consequence of metamorphosis by contact with igneous masses. It is also found in serpentine schists. Extremely beautiful specimens come from the saccaroid dolomia of Campolongo and Val Tremola (Switzerland), Val Devro and Val di Lanzo (Italy), the USSR and USA. It is also called anphibolic asbestos and is used in the same way and for the same purposes as asbestos.

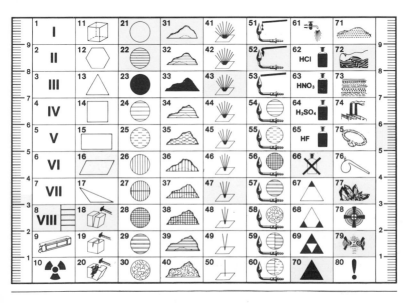

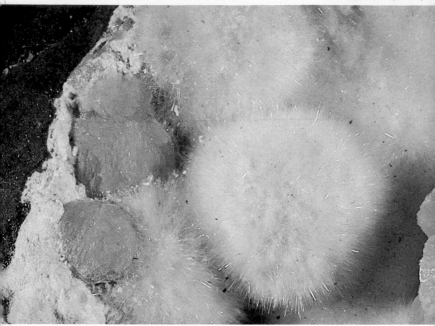

$Ca_{1.5}[Si_3O_6(OH)_3].1.5H_2O$ Poona (India)

Hydrated calcium silicate. A rare mineral belonging to the tobermorite group, which also includes other rare varieties: riversideite, plombierite, nekoite and scawtite. It was first found on Disko island (Greenland) and was studied by W. Kobell (1828). It occurs as globular formations, 2–3cm in diameter, consisting of acicular crystals which can be white, bluish, yellowish with pearly hues, soft and very showy. It can also be massive (fibrous and compact), and can be confused with mordenite. It is found on the Faroe Islands, in Iceland, Rio Putangang (Chile) and Ronah and Poona (India). The so-called 'okenite' of Nursoak (Greenland) is actually wollastonite.

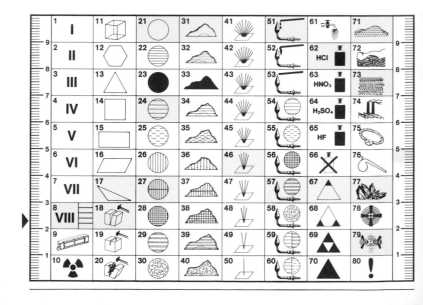

$Ca_2Al^{(6)}[(OH)_2AlSi_3O_{10}]$ Monte Rosa, Piedmont (Italy)

Hydrated calcium aluminium silicate. It occurs as tabular crystals, often in a fan-shaped formation; also as botrioidal, mamillary or stalactitic aggregates. It is a secondary mineral forming in hydrothermal deposits within cavities of volcanic rocks and in the fissures of crystalline schists (gneiss, anphibolites). It is very hard and suitable for making ornaments. Splendid specimens, stalactitic and greenish, have been found at Poona (India); botrioidal masses come from Paterson and Bergen Hill (New Jersey); mamillary formations with cylindrical crystals were found in parts of France and northern Italy.

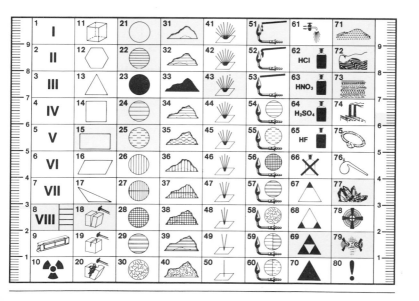

$Mg_3[(OH)_2Si_4O_{10}]$ Val Chisone (Italy)

Hydrated magnesium silicate. Isolated crystals unknown; usually pseudohexagonal scales arranged in rosettes or as scaly and foliated masses. The most compact variety is called steatite. Feels greasy. It is one of the silicates most widely used in industry: paper, textiles, cosmetics, dyes, pharmaceutical products, explosives, refractory and insulating materials (when mixed with felspar and clay and baked at 1480°C). Important deposits have been found in the Pyrenees, Styria (Austria), the Urals, Korea, India, Transvaal (S. Africa), Canada, and many American states. The deposits of northern Italy and Sardinia are also highly productive.

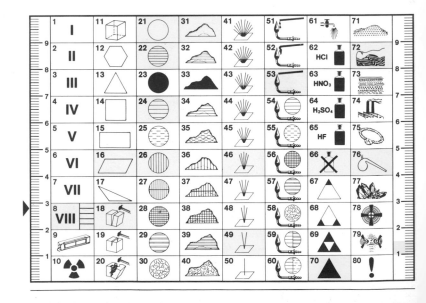

$Mg_6[(OH)_8Si_4O_{10}]$ Val di Susa (Italy)

Hydrated magnesium silicate. Serpentine, a magnesium phyllosilicate, is mainly made up from scaly antigorite and fibrous chrysotile. These fibres derive from furled-up antigorite scales which means that chrysotile, notwithstanding its fibrous appearance, is actually a phyllosilicate. It can be found in three polymorphic varieties: clino-chrysotile and monoclinic; orthochrysotile and parachrysotile; and rhombic. Known as asbestos when the fibres are soft. Large deposits are found in the serpentines of the Alps: Val Malenco (Sondrio), where the fibres can be up to 2m long; Balangero (Turin), with short fibres, ideal for the production of asbestos lumber; Val di Susa, Val d'Aosta and Val Sesia (Italy). Equally important deposits exist in Canada, the USA, the USSR and Zimbabwe.

Na(AlSi$_2$O$_6$).H$_2$O   Val di Fassa (Italy)

Hydrated sodium aluminium silicate. A zeolitic mineral belonging to the leucite group. It occurs as well formed, icositetrahedral crystals, rarely cubic. It is milky white, pink or even greenish. Very beautiful colourless crystals have been found on Cyclops Island, north of Sicily, those from the Alpe di Siusi (Bolzano) are milky or pink and can be up to 6–7cm in diameter. Smaller are the crystals found in Val di Fassa (Trento), in the basalts near Vicenza and in Val dei Zuccanti (Italy). Excellent specimens come from Australia, Bergen Hill and Paterson (New Jersey).

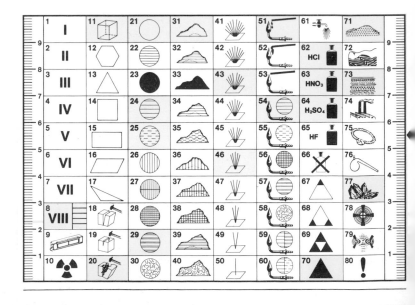

$K(AlSi_2O_6)$ Roccamonfina (Italy)

Potassium aluminosilicate. The best known of felspars, it is the dimorphic variety which exists in cubic form as leucite β above 605°C; at a lower temperature, its structure is tetragonal with external cubic appearance, and it is called leucite α. Its crystals, called 'leucitohedrons', are pseudo-icositetrahedrons. It is a typical mineral of the alcaline, igneous rocks, with low levels of silicon. Central Italy, from Lake Bolsena to Mount Vesuvius, is rich in such deposits and offers beautiful specimens of leucite, either within their ore or free. Leucites are also present in the basalts of Asia, Germany, Montana and Wyoming (USA), the Congo and Australia.

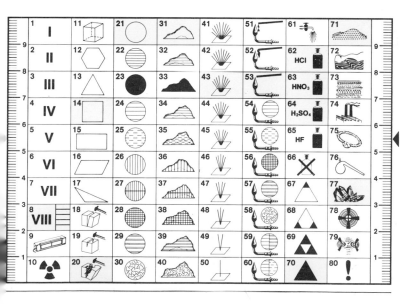

K(AlSi$_3$O$_8$)   Val di Susa (Italy)

Tectosilicate of aluminium and potassium. This is the almost transparent variety of orthoclase, typical of the Alps, where it is frequently found in the fissures of granites and gneiss. Its monoclinic crystals are quite large, up to 25cm on the longer edges; its colour is whitish, greenish or semi-transparent. The opalescent variety found in Sri Lanka and Burma is called moonstone and is used by jewellers. The discoveries in the Swiss Alps (St Gotthard, Val Medel, Val Cristallina) and in Austria (Zillertal) are famous; excellent samples come from Valle Aurina and Val di Vizze (Italy); Val di Susa (Piedmont) has recently yielded fabulous specimens.

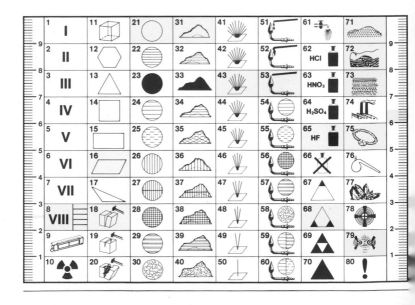

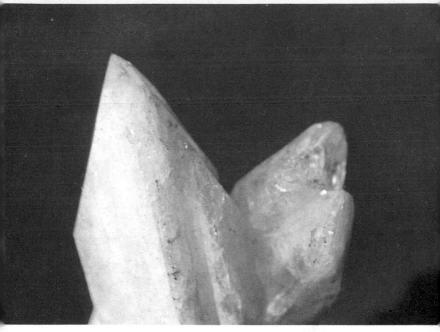

$Ca(B_2Si_2O_8)$ S. Louis Potosi (Mexico)

Calcium boron silicate. A rare mineral which crystallizes in rhombic prisms with a pattern similar to that of topaz. It is colourless or yellowish-white, or even brownish. It is typical of fissures, and can be found in the Alpine lithoclases where crystals are no longer than a centimetre but extremely clear and many-faceted. It was first found at Danbury (Connecticut); excellent samples later came from other localities: S. Louis Potosi, Maharitra (Malagasy Republic), Obira (Japan) and Bolivia. The most famous alpine area is the Grisons (Switzerland: Piz Vallatscha, Piz Mies), where danburite often encrusts adularia crystals.

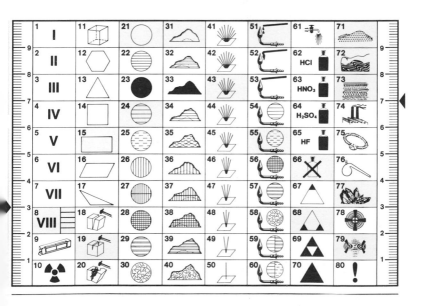

$Na_2(Al_2Si_3O_{10}).2H_2O$ Poona (India)

Hydrated sodium aluminium silicate. A zeolite deriving from the alteration of
nepheline, sodalite or plagioclase, and usually as secondary formation in basic eruptive
rocks. Its rhombic crystals, sometimes huge (at Asbestos, Canada, up to 1m
long × 15cm wide), are usually thin, elongated and grouped in radiating clusters or
spheres. They are colourless, yellowish or pinkish due to inclusions. Fine samples came
from the lava of Puy de Dome and Puy de Marmant (France), Aussig and Salesl
(Czechoslovakia), from the basalts of N. Ireland (Antrim), and Skye. Superb specimens
have been found in India, Canada, the USA, Brazil, and good ones in the basalts of
northern Italy and Sicily.

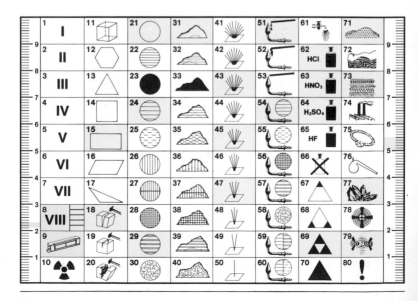

$Ca(Al_2Si_3O_{10}).3H_2O$ Poona (India)

Hydrated calcium aluminium silicate. A zeolite analogous to natrolite but containing calcium instead of sodium. Present not only within basic eruptive rocks but also in lithoclases of gneiss, sienite, and metamorphosed limestones. When heated it inflates and forms spirals (hence its name, meaning 'worm' in Greek). Its crystals are splendid monoclinic prisms, a few centimetres long, colourless or white, often in clusters. It is found in Switzerland (Val Giuv, Val Calanca) and Austria (Obersulzbachtal), as well as in the basalts of Iceland, of the Faroe Islands, Scotland and Eritrea (Ethiopia). Excellent specimens came from Poona, Rio Grande do Sul (Brazil) and Table Mountain (Colorado). Massive forms exist in northern Italy and on Mount Vesuvius.

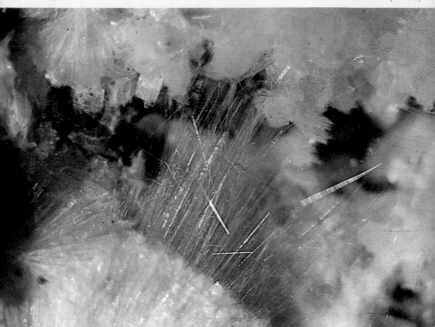

$Na_2Ca_2(Al_2Si_3O_{10})_3.8H_2O$ Oregon (USA)

Hydrated sodium calcium aluminium silicate. A zeolite with a chemical composition half way between natrolite and scolecite. It crystallizes in monoclinic, prismatic crystals, elongated, often growing with natrolite, scolecite or thomsonite, and gathered in tufts or clusters. It is white, grey, tending to yellow. It occurs in fairly large quantities in Northern Ireland, the Faroe Islands and Talisker Bay (Scotland). It can also be found in the andesites of Mount Imeretin (USSR), in Siberia, in the basalts of Grant County (Oregon) and in Sardinia.

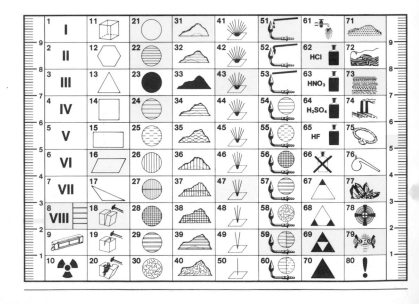

As_2S_3 Quiruvilca (Peru)

Arsenic sulphide. The name is derived from the latin *auropigmentum* meaning gold pigment, because this mineral was widely used, particularly in the Orient, as a golden yellow colourant. Crystals are very rare; only recently, fine specimens have been found at Quiruvilca which present prismatic monoclinic crystals of 3 × 2cm and more. The colour varies from lemon yellow to golden yellow, orange yellow and brownish yellow. It forms through sublimation within effusive rocks and through the alteration of realgar (As_4S_4). It is found in quantity in Kurdistan, Macedonia (Greece) and Rumania, in Iran, Georgia (USSR) and Utah (USA). Smaller quantities can be found in E. Germany, on the Alps, and on Mount Vesuvius.

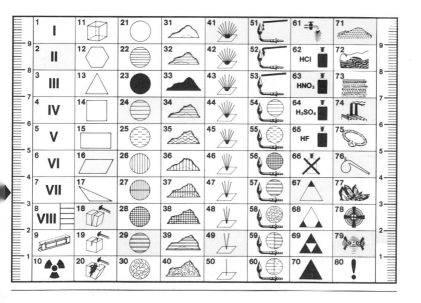

$CaMg(CO_3)_2$ Arkansas (USA)

Calcium magnesium bicarbonate. It occurs in rhombohedrons with curved faces, or 'saddle-shape rhombohedrons'; the pink colour is due to cations which take the place of a minimal amount of magnesium. If bivalent iron prevails over magnesium one has the variety ankerite, while the rare variety kutnahorite is produced by a prevalence of manganese on magnesium. Localities now famous for their excellent dolomite are Joplin in Missouri, Roxbury in Vermont, Hoboken in New Jersey, Stony Point in North Carolina and Warsaw in Illinois, while splendidly clear, jewel-like crystals are found in the saccaroid dolomias of northern Italy (Novara) and Switzerland (Campolongo and Binnthal).

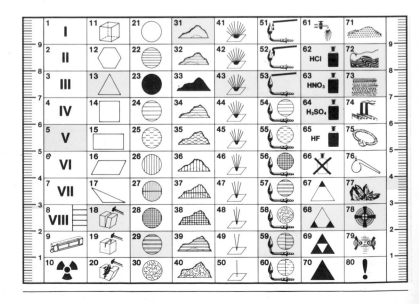

$Pb(CrO_4)$ Dundas (Tasmania)

Lead chromate. A very beautiful and rare mineral, much sought after by collectors, and known in the past as 'Siberian red lead' owing to its deep orangey-yellow colour and its provenance. It was first found at Beresov (Urals). It occurs as separate, prismatic and monoclinic crystals, up to 15cm long, finely striated and growing in rather disorderly groups. It forms in the oxidation zone of lead deposits, particularly in the presence of eruptive rocks rich in chromium. The best samples come from Dundas, Goyabeira (Minas Gerais, Brazil), Mursinsk and Nizhne-Tagilsk (Urals), Luzon (Philippines), Nontron (France), California and Arizona (USA).

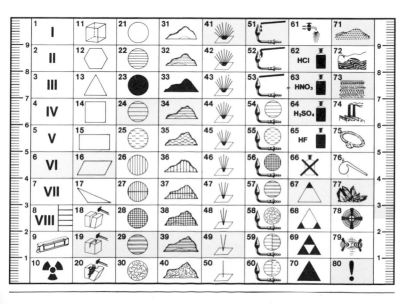

$Mn_3Al_2(SiO_4)_3$ Praborna, Val d'Aosta (Italy)

Manganese aluminium nesosilicate. A valuable garnet, brownish-yellow, honey-yellow, orange or reddish, it can contain small quantities of bivalent and trivalent iron as substitutes of manganese and aluminium. It is found in granite, pegmatite and metamorphic rocks rich in manganese ores. It can be found at Aschaffenburg, in Spessart (Bavaria; hence its name); in the pegmatites of Ansirabe (Malagasy Republic), which yield transparent, deep-yellow crystals; in the granites of Amelia Court House (Virginia) and Sierra San Pedro (Mexico), and in the granites of the Piedmontese valleys (together with piemontite) and Elba.

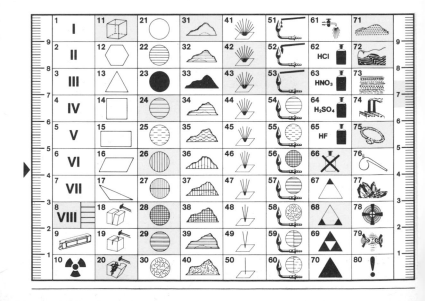

$Al_2(F,OH)_2SiO_4$ Minas Gerais (Brazil)

Basic aluminium fluorosilicate. It occurs as very beautiful rhombic crystals, thick in habit, sometimes gigantic (over 250kg). It is of hardness 8 on the Mohs scale and pyro-electric. Its colour varies from brownish yellow to blue, green, violet and reddish; the limpid, colourless specimens are excellent. When set in sand and heated to 300–450°C it becomes pinkish-brown and is called 'burnt topaz'. The most famous samples come from Minas Gerais (Fazenda do Funil) and from the mines of Ouro Petro. Natural 'burnt' topazes have come from Sankara (Siberia), splendid blue ones from Murinska (Urals) and yellow ones from Burma. Small specimens can be found in the granites of Elba.

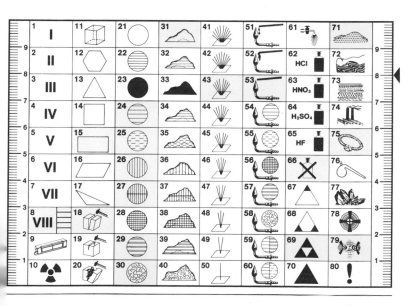

$Ca_2Al_3[O|OH|SiO_4|Si_2O_7]$ Val di Susa (Italy)

Basic calcium aluminium silicate. It is the aluminium-ore member of the epidote series, which includes minerals constituted by isomorphic compounds of clinozoisite and epidote (or pistacite), which contain iron. Clinozoisite occurs as tabular crystals or compact, granular or acicular masses. It is sometimes colourless, but usually white, greyish, greenish-yellow or attractively pink. It is associated with crystalline schists, contact rocks, and eruptive rocks through the alteration of plagioclases. Fine samples come from Switzerland, Austria (Zillertal), Malagasy Republic, Brazil, and the Piedmontese Alps (Italy), particularly Val di Susa which recently yielded splendid specimens.

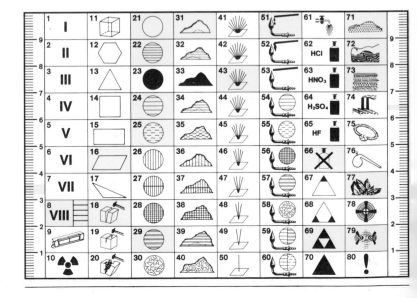

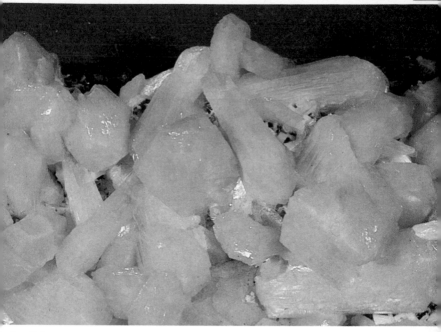

$Ca(Al_2Si_7O_{18}).7H_2O$

Poona (India)

Hydrated calcium aluminosilicate. Also called desmine, it is a zeolite of the heulandite group. It is radially or spherically massive, or occurs as irregular crystals growing in sheaf-like aggregates, or even tabular, idiomorphic crystals, often twinned in a cross with simulated rhombic symmetry. It is colourless, white, grey, yellowish or pinkish-brown. It forms in the cavities of basalts and within the lithoclases of gneiss and granite. It is found on Skye and at Kilpatrick (Scotland), in the Faroe Islands, in Nova Scotia (Canada), New Jersey (USA), Rio Grande do Sul (Brazil), Poona, and the alpine valleys near Trento and Bolzano (Italy)—associated with the variety puflerite which occurs in transparent, radial spheres—as well as parts of the Apennines and Sardinia.

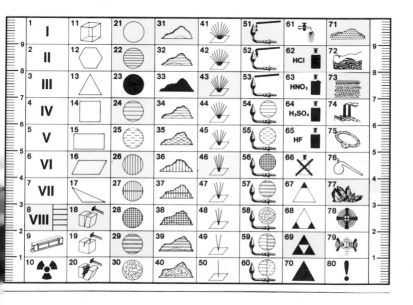

$(Na_2Ca)(Al_2Si_4O_{12}).6H_2O$ Victoria (Australia)

Hydrated sodium calcium aluminosilicate. A very attractive zeolite closely analogous with cabasite, so much so that when first found it was thought to be a sodium cabasite; later, its structure revealed it to be a new variety. It crystallizes in the hexagonal system in usually idiomorphic and frequently twinned crystals; also in radiating aggregates. It is colourless, yellowish, pink, or brightly orange. It can contain small percentages of potassium. It is found at Glenorm and Magee (nr. Antrim, N. Ireland), Berghen Hill and Cap Blomidon (Nova Scotia, Canada), Skye and the Crimea. Excellent pink specimens have come from the basalts of Montecchio Maggiore (nr. Vicenza, Italy), with crystal faces of about 1cm.

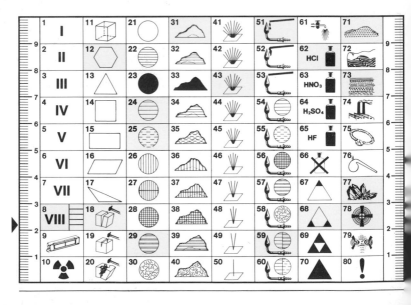

Cu   Lake Superior (USA)

Native copper. Rarely found as cubes, octahedrons, tetracisesahedrons, it is more common in compact, fibrous or denritic masses. Its characteristic colour is called 'copper red', but easily oxidizes when exposed to the air, becomes dark and carbonated into green hues. It forms in the oxidation area of sulphide deposits where hydrothermal solutions of copper salts are acted upon either by organic substances or by rocks containing magnetite. Native copper is widespread but only occasionally found in large quantities. The richest area, by now nearly worked out, is the Keweenaw peninsula on Lake Superior, where the best crystallized samples have come from together with blocks weighing up to 420 tons. Also famous are the deposits at Huelva (Spain), Corocoro (Bolivia), Oneta (Mexico), Broken Hill (Australia), Chile, Zambia and the USSR.

$CaTiO_3$ Val di Susa (Italy)

Calcium titanium oxide. It occurs in splendid pseudocubic crystals, vertically striated, coloured yellow, brown, grey and black. Rare soils and niobium can be included in the darker specimens, thus originating the varieties dysanalyte, knopite and loparite. Perowskite is an accessory in eruptive rocks and splendid specimens can be found both in chlorine and talc schists and within serpentines in contact with limestones. Famous samples have been found at Alno (Sweden), the Urals, Oka (Canada), Magnet Cove (Arkansas), as well as Italy: Sondrio, Bolzano, Val d'Aosta, Val d'Ala and, recently, in great quantity, Val di Susa.

$(U,Ca)_2(Ti,Nb,Ta)O_6(O,OH,F)$ Betafo (Malagasy Republic)

Complex uranium calcium titanium niobium oxide. An extremely rare radioactive mineral, it usually occurs in octahedral crystals, less frequently as compact masses. Its formula is not well-defined; its colour is black turning to brownish yellow through surface alteration. Its uranium content varies between 15 per cent and 25 per cent. It is often associated with other complex oxides such as fergusonite and euxenite. The best-known sources are Malagasy Republic (Betafo, Ampangabe, Samiresy), whence the best crystals come, Lake Baikal (USSR), Tangen (Norway) which yields unadulterated betafite, Ontario, and, recently, the pegmatites of Cosasca (nr. Novara, Italy).

α-TiO₂   Valle d'Oropa (Italy)

Titanium dioxide. With ilmenite it is the most common titanium ore. It occurs in various shapes and colours: long, thin needles which, in quartz, are called 'Venus hair'; or elongated, columnar prisms, often striated and twinned at right angles; its colour is generally red (l. *rossiccio* = reddish), but it can be in various hues of yellow, brown and black. It is used to extract titanium, if found in worthwhile quantities. Fine crystals can be found in the Alps: Switzerland (Grisons, St Gotthard, Binnthal), Austria (Rauris, Zillertal, Sau Alpen), Italy (Val di Vizze and Valle Aurina near Bolzano, Alpe Devero near Novara, and recently Valle d'Oropa and Val d'Aosta) where the deposits of Pian Paludo (Savona) are large enough to be exploited.

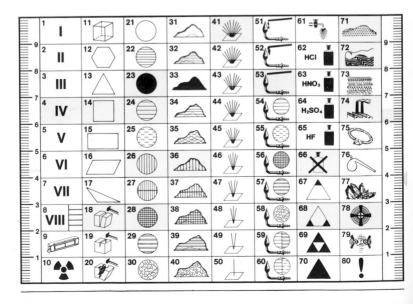

$FeCO_3$ Traversella (Italy)

Iron carbonate. One of the best iron ores, it contains up to 48.3 per cent iron. It occurs in rhombohedral crystals, whitish when fresh but turning to brown through oxidation of the iron. It is usually found in scaly masses; also radially fibrous or cryptocrystalline. Soluble in hydrochloric acid when heated. Important vein deposits have been found at Bilbao (Spain), Ivigtut (Greenland), Huttenberg and Herzberg (Austria), Tunisia and Algeria. Large sedimentary deposits occur in Germany, France (Lorraine), Rumania, southern England, Scotland and Pennsylvania. Famous crystals have been found in Styria (Austria), Cornwall (England), Canada, Brazil, Tuscany and Sardinia (Italy).

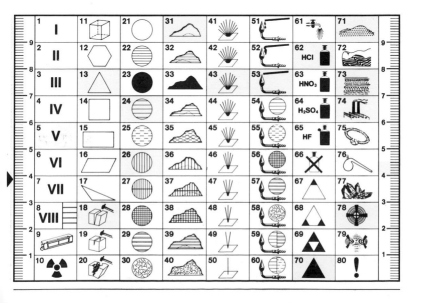

Ba(SO₄)   Barega, Sardinia

Barium sulphate. A fairly common mineral often associated, as gangue, with low-temperature, hydrothermal veins of galena, blende, siderite, antimonite, fluorite, dolomite and quartz. Its very beautiful crystals are prismatic or tabular, but it can also be massive, lamellar, spathic or concretionary. Its colouring is varied and attractive: brown, yellow, blue, green and red; some barites can be colourless and extremely limpid, others whitish. Excellent samples come from Rumania (Felsobanya), from Cornwall, Cumbria and Derbyshire (England), from Freiberg (E. Germany), Pribram (Czechoslovakia), and Italy (Sardinia; near Turin; in the sandstones near Leghorn). Beautiful 'desert roses' formations can be found in the Sahara and in Oklahoma (USA).

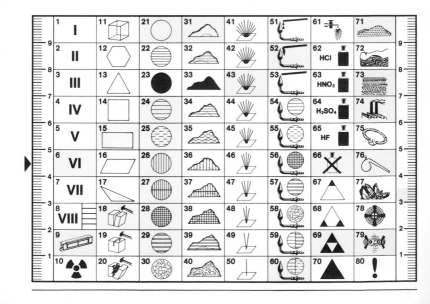

Ca(WO₄)   Traversella (Italy)

Calcium tungstate. An interesting mineral which occurs in splendid crystals of various colours: yellow, reddish-grey, greenish-brown. It normally contains small quantities of molybdenum and rare soils. Very sought-after by collectors, it is used in industry to extract tungsten. It is luminescent to ultra-violet rays of short frequency and emits pale blue light. The best European specimens have come from the mines of Traversella, where they were found in chloritic-talcous rocks: average dimensions were 3–4cm and up to 10cm in the largest samples. It is also found in other parts of Italy (Val di Fiemme, Val Sugana), in Brazil, Bolivia, Burma, Japan, Korea, China and the USA.

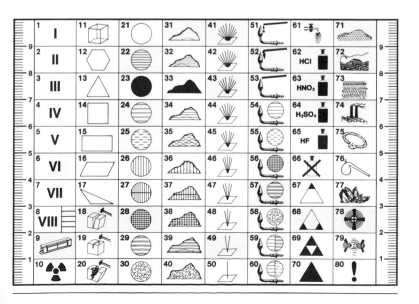

Zr(SiO₄)　　　　　　　　　　　　　　　　　　　　　　Malagasy Republic

Zirconium silicate. It occurs as tetragonal, bipyramidal crystals, sometimes elongated and prismatic, ending in a bipyramid. It often includes hafnium, thorium and rare soils. It is found in primary deposits in such eruptive rocks as granite, pegmatite, sienite, and in secondary alluvial deposits. Zircon is used to extract the oxide (ZrO_2) which, since it fuses at 3000°C, is employed in the manufacture of refractory materials; zirconium, used in the iron and steel industry, and for nuclear reactors; and hafnium, for missiles and Roentgen rays. Zircon is widely mined in the alluvial deposits of Sri Lanka, Burma, Thailand, Brazil, Malagasy Republic and Australia. Primary deposits have been found in Norway, Sweden and the Urals. Small quantities are also recorded in the Alps.

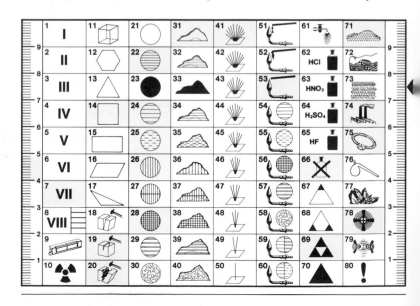

$AlFe_2O_3(OH).4Al_2(O|SiO_4)$ Georgia (USA)

Iron aluminium silicate. It is a widely-found mineral, well known for its cross-like twinnings. It crystallizes in rhombic prisms, short and brownish-red or blackish. The twins are in the shape of St Andrew's cross if the angle between the arms of the two prisms is 60°; of a Greek cross if the angle is 90°. Staurolite is a frequent accessory mineral in crystalline schists with garnets, cyanite and tourmaline. It is also present in alluvial deposits, being resistant to physical and chemical agents. Large crystals come from Goldenstein (Czechoslovakia), Aschaffenburg (Bavaria), Brittany and various localities in the USA. Clear and famous crystals are found at Pizzo Forno (Switzerland) and are abundant in the mica schists of Lake Como, Lombardy and Trentino (Italy).

$(Ca,Na)_2(Al,Mg)[(SiAl)_2O_7]$ Marino (Italy)

Calcium sodium magnesium aluminosilicate. The most important member in the series of its name, which comprises akermanite, gehlenite, gugiaite and hardystonite. Tetragonal, it crystallizes in prisms flattened on the basal pinacoid, or slightly columnar; usually yellow, brownish-red or grey but also colourless. It is found in the more recent basic effusive rocks of the leucite or nepheline type, occasionally as the main component. Famous basalts where melilite can be found with augite, olivine and perowskite are those in Germany, Czechoslovakia, Kola Peninsula (USSR) and Colorado (USA); it is also found in the lavas around Rome, Orvieto and Mount Vesuvius (Italy).

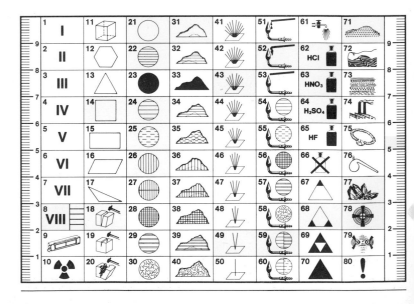

K(AlSi₃O₈)   Baveno (Italy)

$K(AlSi_3O_8)$

Potassium aluminium tectosilicate. Monoclinic, it occurs in prismatic crystals, either elongated, tabular or flattened, very often twinned through compenetration (Charlsbad law) or contact (Manebach and Baveno laws). It is colourless or grey, yellow, blue and pink. Several fine specimens have come from Baveno, lesser ones from Varese and Elba. Grey orthoclase is found at Villadreu (Spain), blue samples at Lake Baikal (USSR), clear yellow ones in Malagasy Republic, red ones in Scotland, Switzerland, Germany and various places in Siberia. It has various applications both in jewellery and industry: ceramics, glass, detergents.

$K(AlSi_3O_8)$ Zacatecas (Mexico)

Potassium aluminium tectosilicate. It represents the triclinic modification, at low temperature, of the compound $K(AlSi_3O_8)$ which crystallizes also as orthoclase and sanidine. Its white, grey, yellowish, rose or more rarely green crystals resemble orthoclase, and are subjected to the same twinning laws as well as a double polysynthetic twinning which gives it a grate-like structure. The blue-green variety called amazonite is used in jewellery, when it is cut as 'cabochon' and polished. Microcline is a typical component of granites and gneiss. Good samples are found in the granites of the Isola della Maddalena (Italy), the Urals and in various places in the USA and Canada. Amazonite is mined in Malagasy Republic, Norway, Brazil, India, Canada and the USA (Colorado and Virginia).

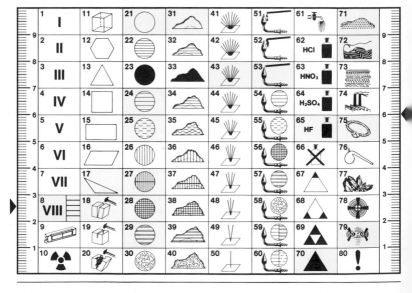

GLOSSARY

Accessory: a mineral usually associated with a given rock but always in small quantities.

Accidental: a mineral which can only occasionally be found in a given rock, and which usually does not appear at all.

Acicular: a thin, rather elongated crystal.

Aggregate: a group of crystals of the same or of different kinds.

Allochromatic: a mineral assuming various colours.

Allotigenous: a mineral which is usually found in a different location from the one where it formed (alluvial gold).

Alpha particle: a particle consisting of two protons and two neutrons, produced by the radio-active decay of certain nuclei.

Amorphous: a state of solid matter lacking any order in the distribution of particles.

Analyser: part of a reflection goniometer to which the eye is applied, or the part containing nicol of a mineralogist's microscope.

Anion: atom or atomic group with negative charge.

Anisotopy: the phenomenon according to which the physical properties of a mineral vary according to its direction.

Arnion: spherical or kidney-shaped crystal aggregate.

Atom: the smallest particle of an element that enters into the composition of that element and its compounds.

Autigenous: a mineral which is discovered in the place where it formed.

Bacillar: a very elongated, thin crystal.

Beta particle: a positive or negative electron produced by the nucleus during certain types of radioactive disintegration.

Botryoidal: globular crystal aggregate.

Bunsen: a blow-pipe used to burn gas.

Cabochon: a style of gem-cutting which produces a more or less convex surface.

Cation: ion with a positive charge

Chatoyancy: the light, reflected by certain minerals, which concentrates along thin, iridescent and pale bands

Cleavage: the ability of crystals to break along certain planes which are equivalent to planes of least cohesion within the lattice.

Columnar: elongated, but not very thin, crystals.

Conchoidal: a cleavage with curved surfaces, such as in broken glass.

Concretion: deposits of micro-crystals on a bed.

Counter: an instrument used to detect particles emitted by radioactive substances.

Crystalline form: the group of crystal faces which are physically equivalent.

Curie: a measurement unit for radioactivity which is equivalent to 3.7×10^{10} disintegrations per second in any given nuclide.

Density: absolute = the weight of the volume unit of a body; relative = the ratio between the weight of the body and that of an equal volume of distilled water at $4°C$.

Density bottle: an instrument used to measure a mineral's specific gravity.

Deposit: an economically useful concentration of a mineral.

Diamagnetic: minerals which are mildly repelled by a magnet.

Druse: a group of crystals growing on a flat matrix.

Elementary cell: the smallest portion of a crystalline lattice which is characterised by the crystal's properties.

Essential: minerals which are part of, and qualify, a rock.

Evaporites: sedimentary rocks with chemical origin which formed through the evaporation of water.

Femic: rocks containing certain amounts of silica (44 − 52 per cent).

Ferromagnetic: minerals which are strongly attracted by a magnet.

Fluorescence: phenomenon of luminescence which is interrupted by the removal of the exciting radiation.

Formula: chemical composition of an element or compound expressed with symbols.

Fracture: break along irregular surfaces.

Fumarole: the emission of gases and vapours following a volcanic eruption.

Gamma ray: a highly penetrating electromagnetic radiation emitted by the nucleus of certain radioactive substances during their decay; similar in nature to X-rays.

Gangue: the useless part of mineral deposits, usually consisting of rocks or minerals associated with the useful one.

Geminate: the association, according to certain laws, between one or more crystals of the same kind.

Geodes: a spherical or sub-spherical cavity lined with crystals.

Habit: the geometrical shape, consisting of one or more forms, which is normally assumed by the mineral.

Hardness: a mineral's resistance to scratching.

Holohedral: in a system, the class containing the largest number of symmetry elements.

Hydrothermal: mineral deposits formed within the fissures of rocks and derived from warm watery solutions of magmatic origin.

Hyposiliceous: rocks containing less than 52 per cent of silica (SiO_2) whether free or as part of silicates.

Idiochromatic: a mineral which never changes its colour.

Igneous: rocks originated by the solidification of magma.

Inclusions: either solid or liquid or gaseous particles contained in a mineral.

Ion: charged atom or atomic group.

Isomorphism: chemically different minerals which adopt the same crystalline structure.

Isotropes: bodies whose physical properties are equal in all directions.

Kobell scale: scale of mineral fusibility.

Lamellar: crystals developing mainly in two directions.

Lithoclase: cavity or fissure in a rock, usually lined with crystals.

Luminescence: emission of light by a mineral exposed to exciting radiation.

Lustre: the intensity of light reflected from the surface of a crystal.

Magma: rocks in their viscous fluid state.

Magmatic segregation: the separation of minerals from magma during the first phase of solidification.

Mamelon: rounded crystalline aggregate.

Metamorphic: a rock which has varied its structure under certain conditions (pressure, temperature, contact with magma).

Metasomatosis: the variations occurring in a rock due to the addition of some minerals and the subtraction of others.

Meteorite: interplanetary body landing on the surface of the Earth.

Mohs scale: scale of mineral hardness.

Molecule: the smallest quantity of an element or compound with their respective characteristics and existing as a free entity.

Ochre: earthy mass comprised of minute crystals.

Outcrop: alteration zone of ore deposits which appears on the surface of the ground.

Oxidation: the effect of oxygen on minerals, which causes the formation of new species.

Paramorphosis: the change from one structure to another, within the same substance, without any alteration of the crystalline form.

Pepita: small spherical metallic mass found in alluvial deposits.

Persiliceous: rocks containing over 65 per cent of silica.

Phosphorescence: phenomenon of luminescence which is not interrupted by the removal of existing radiation.

Piezo-electricity: the phenomenon by which certain crystals, subjected to pressure, become electrically charged.

Pneumatolitic: the vein formed by the consolidation of gaseous magma.

Polariscope: instrument used to obtain polarized light.

Polymorphism: the phenomenon by which the same mineral can show more than one lattice structure according to the conditions under which it was formed.

Projectile: a solid body ejected by a volcano.

Pseudomorphism: when a mineral maintains its form but is substituted by another mineral.

Pyro-electricity: the phenomenon by which certain crystals, subjected to heat, become electrically charged.

Radioactivity: spontaneous nuclear disintegration with emission of particles or electromagnetic radiation, due to the instability of certain nuclei. Measured in curies.

Reduction: the chemical process whereby oxygen is extracted from minerals.

Scalars: physical quantities which do not vary with direction.

Schistosity: typical structure of metamorphic rocks whose mineral components are arranged in parallel or sub-parallel planes.

Secondary: a mineral formed by the alteration of another.

Sedimentary: rocks derived from igneous or metamorphic ones and deposited into strata by water.

Spatic: a mineral with extremely easy cleavage.

Specific weight: *see* Density, relative

Sublimation: the direct transition from gaseous state to solid state without going through the liquid state.

Tabular: flattened crystals showing a marked development of the pinacoidal faces.

Tenacity: a mineral's resistance to breakage.

Vectors: physical quantities which do vary with direction.

Vein: a magmatic intrusion considerably extended in one direction.

FURTHER READING

BATEMAN, A. M., *Formation of Mineral Deposits*, Wiley (1951)

BATTEY, M. H., *Mineralogy for Students*, Longman (1975)

BAUER, J., *Field Guide to Minerals, Rocks and Precious Stones*, Octopus (1974)

BOEGEL, H., *Collector's Guide to Minerals and Gemstones*, Ed Sinkankas, J., Thames & Hudson (1971)

CLARK, C., *Minerals*, Hamlyn (1979)

CORRENS, C. W., *Introduction to Mineralogy*, Spring Verlag Berlin (1969)

DANA, E. S., *A Text Book of Mineralogy*, Wiley (1954)

– *Minerals and How to Study Them*, Wiley (1949)

DEESON, A. F. L. (Ed), *The Collector's Encyclopedia of Rocks & Minerals*, David & Charles (1973)

EMBREY, P. G & FULLER, J. P. (Ed), *Manual of New Mineral Names*, OUP (1980)

FRYE, K., *Modern Mineralogy*, Prentice Hall (1974)

GREY, R. P. & LETTSOM, W. G., *Manual of the Mineralogy of Great Britain & Ireland*, Lapidary Pubns (1977)

KIRKALDY, J. F., *Minerals and Rocks in Colour*, Blandford Press (1963)

LYE, K., *Minerals and Rocks*, Ward Lock (1979)

MICHELE, V. de, *World of Minerals*, Orbis (1976)

O'DONOGHUE, M. (Ed), *Encyclopedia of Minerals and Gemstones*, Orbis (1976)

PEARL, R. M., *How to Know Minerals and Rocks*, McGraw (1963)

PHILLIPS, W. J. & PHILLIPS, N., *Introduction to Mineralogy for Geologists*, Wiley (1980)

POUGH, F. H., *A Field Guide to Rocks and Minerals*, Boston (1960)

ROBERTS, W. L., RAPP G. and WIEBER J., *Encyclopedia of Minerals*, Van Nostrand Co (1974)

SCHUMANN, W., *Gemstones of the World*, N.A.G.P. (1977)

– *Minerals and Rocks*, Chatto (1978)

SINKANSAS, J., *Mineralogy for Amateurs*, Ed Van Nostrand Co (1964)

WATSON, J., *Rocks and Minerals*, Allen & Unwin (1979)

ALPHABETICAL INDEX

SYSTEMATIC INDEX

Numbers given refer to colour plates

WILD HERBS: A FIELD GUIDE
Jacques de Sloover & Martine Goosens

Whether used as a practical identification guide in the
field, or for armchair browsing, this book offers a great
deal of information succinctly presented. The herbs in the
144 stunning colour plates are grouped by colour. To
identify a plant, simply open the guide at the pages
bordered by the colour corresponding to that of the flower
and you will soon find the plant itself. This ingenious
system is time-saving, and the close proximity of
illustrations of similarly-coloured flowers helps to avoid
misidentification.

For the purposes of this book, a herb is defined as a
useful plant, one which is used to cure, to feed, to flavour
dishes, to dye wool, or for any other specific aim.
Pictograms presented alongside each colour plate
summarize other properties – aromatic, medicinal and
culinary, which parts are efficacious, when the herb is at
its prime, where it grows, when it flowers. This at-a-
glance information is supplemented by useful appendices,
a glossary and notes on further reading.

*210×120mm (8¼×4¾ in) 144 colour plates, 144
pictograms*

MUSHROOMS & TOADSTOOLS: A COLOUR FIELD GUIDE

U. Nonis

Anyone interested in collecting mushrooms, whether to study them scientifically or simply to enrich their everyday diet with their nutritional value, will find this book an invaluable guide. It is based on a new descriptive system: each of 168 colour photographs shows specimens in their natural habitat and is accompanied by a pictogram giving identification – at a glance – of their principal characteristics, combining a wealth of information with simplicity of presentation. Identification is further aided by the colours on the margins of the pages which reflect those of the fungi.

An introduction describes the main genera, their habitat, dangerous or valuable properties, and directions for collecting and growing them. Further Reading, Etymology of Scientific Terms and indexes contribute to the unique value of this guide.

210×120mm (8¼×4¾in) 168 colour plates, 79 line illustrations, 168 pictograms

MOUNTAIN FLOWERS: A COLOUR FIELD GUIDE
S. Stefenelli

Thousands of enchanting flowers grow on the mountain slopes of Europe, and this book will prove an informative and useful guide for those wishing to discover more about them, while appreciating their beauty and understanding the need for their conservation.

Recognition is easy with the aid of the 168 splendid colour photographs. To identify a flower, simply match the colour of the flower to the corresponding colour section and your task becomes easy. Once it has been clearly identified, the pictograms which accompany the plates will enable the beginner and the serious botanist to discover at a glance all the other interesting facts about the flower. A bookmark showing the key to the pictogram, a section on habitat, a glossary of pharmaceutical terminology, a bibliography and two indexes add to the value of the book.

210×120mm (8¼×4¾in) 172 colour plates, 168 pictograms

Colour of the powder

31	White or pale
32	Grey
33	Black
34	Yellow
35	Green
36	Rusty brown
37	Blue
38	Red
39	Orange, pink
40	Violet, purple

LUSTRE

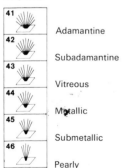

41	Adamantine
42	Subadamantine
43	Vitreous
44	Metallic
45	Submetallic
46	Pearly
47	Silky
48	Resinous
49	Greasy, waxy, oily
50	Dull

Fusibility

51	Fusible
52	Resistant to fusion
53	Non-fusible

Colour in a flame

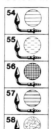

54	Yellow
55	Green
56	Red
57	Orange
58	Violet
59	Yellow-orange
60	Yellow-green

Solubility

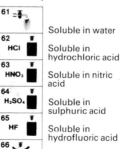

61	Soluble in water
62 HCl	Soluble in hydrochloric acid
63 HNO₃	Soluble in nitric acid
64 H₂SO₄	Soluble in sulphuric acid
65 HF	Soluble in hydrofluoric acid
66	Insoluble in acids

Occurrence

67	Very rare
68	Rare
69	Common
70	Very common

Origin

71	In igneous rocks
72	In metamorphic rocks
73	In sedimentary rocks and hydrothermal deposits

Use

74	Metallurgy, building industry, etc
75	Jewellery
76	Chemical industries
77	Private collectors, scientific research

Polarised light with converging nicols

78	Interference figure of dimetric crystals
79	Interference figure of trimetric crystals

Preservation

80	Perishable

KEY TO SYMBOLS USED

Scale of specific gravity

9
8
7
6
5
4
3
2
1

Scale of hardness

9
8
7
6
5
4
3
2
1

Mineral Classes

1 **I**	Native elements
2 **II**	Sulphides
3 **III**	Halogenures and Halides
4 **IV**	Oxides and Hydroxides
5 **V**	Carbonates, Nitrates, Borates
6 **VI**	Sulphates, Chromates, Molybdates, Wolframates
7 **VII**	Phosphates, Arsenates, Vanadates
8 **VIII**	Silicates

Crystal systems

11	Cubic
12	Hexagonal
13	Trigonal
14	Tetragonal
15	Rhombic
16	Monoclinic
17	Triclinic

Colour of light-reflecting minerals

21	White
22	Grey (whether metallic or not)
23	Black
24	Yellow
25	Green
26	Rusty brown
27	Blue
28	Red
29	Orange, pink
30	Violet, purple

Excitation and Radioactivity

| 9 | Fluorescent minerals |
| 10 | Radioactive minerals |

Cleavage

18	Perfect
19	Imperfect
20	Non-existent